Antibiotics and Antibiotic Resistance in the Environment

T0179224

Antibiotics and Antibiotic Resistance in the Environment

Carlos F. Amábile-Cuevas
Fundación Lusara, Mexico City, Mexico

CRC Press
Taylor & Francis Group
Boca Raton London New York Leiden

CRC Press is an imprint of the
Taylor & Francis Group, an **informa** business

A BALKEMA BOOK

Published by: CRC Press/Balkema
P.O. Box 11320, 2301 EH Leiden, The Netherlands
e-mail: Pub.NL@taylorandfrancis.com
www.crcpress.com – www.taylorandfrancis.com

First issued in paperback 2020

CRC Press/Balkema is an imprint of the Taylor & Francis Group, an informa business

© 2016 Taylor & Francis Group, London, UK

Typeset by MPS Limited, Chennai, India

Library of Congress Cataloging-in-Publication Data

Names: Amábile-Cuevas, Carlos F., author.
Title: Antibiotics and antibiotic resistance in the environment / Carlos F.
    Amábile-Cuevas, Fundaciâon Lusara, Mexico City, Mexico.
Description: Leiden, The Netherlands : CRC Press/Balkema is an imprint of the

    Taylor & Francis Group, an informa business, [2016] | Includes
    bibliographical references and index. | Description based on print version
    record and CIP data provided by publisher; resource not viewed.

Identifiers: LCCN 2015045076 (print) | LCCN 2015040185 (ebook) |
    ISBN 9781315679419 (ebook) | ISBN 9781138028395 (Hbk : alk. paper) |
    ISBN 9781315679419 (eBook)
Subjects: LCSH: Antibiotics—Environmental aspects. | Disinfection and
    disinfectants—Environmental aspects. | Drug resistance in microorganisms.
Classification: LCC TD196.D78 (print) | LCC TD196.D78 A434 2016 (ebook) |
    DDC 363.738—dc23

ISBN: 978-0-367-57517-5 (pbk)
ISBN: 978-1-138-02839-5 (hbk)

Table of contents

Preface

Antibiotic resistance is trendy once again. During the last few years, the World Health Organization, and the US and European Centers for Disease Control and Prevention, have all issued reviews on the subject, and stern calls to do something about it. The Infectious Diseases Society of America launched the 10 × '20 initiative, calling for a global commitment to have ten new antibiotics by the year 2020; even the US White House issued a *National Action Plan for Combating Antibiotic-Resistant Bacteria*. The finding of a new antibiotic produced by soil bacteria, something that would have been published in an obscure journal, if at all, had it happened during the last century, ended up as a full paper in *Nature*, and most news outlets worldwide afterwards. Is bacterial resistance to antibiotics something new? Was its emergence unexpected? Is it out of control as never before? Curiously enough, the answer to all three questions is 'no'.

Penicillin, the very first antibiotic, had its major clinical debut in 1942; it was used on the victims of the infamous fire at the "Cocoanut Grove" nightclub in Boston. That very year, Rammelkamp and Maxon reported the development of resistance in *Staphylococcus aureus* after long-term culturing in increasing concentrations of penicillin; and found the same phenotype in four clinical isolates obtained during the course of penicillin treatments (*Exp Biol Med* 51: 386–389). Three years later, Alexander Fleming himself, in his Nobel Lecture, warned that it was "not difficult to make microbes resistant to penicillin in the laboratory", and that "ignorant men" underdosing themselves, would make microbes resistant to the drug. By 1981, antibiotic abuse was so rampant, and antibiotic resistance so common, that Stuart Levy founded the Alliance for the Prudent Use of Antibiotics. Antibiotic resistance made the cover of *Newsweek* in March, 1994 ("Antibiotics – the end of the miracle drugs?") and, half a year later, of *Time* ("Revenge of the killer microbes – are we losing the war against infectious diseases?"). However, nothing spectacular did happen afterwards; mortality rates due to increased bacterial resistance kept rising, at a continuous but not dramatic pace. It wasn't until the end of 2007, when a "18,650" figure at the bottom of a busy table in a *JAMA* paper, created a new boom for resistance. That was the estimated number of yearly deaths caused by methicillin-resistant *S. aureus*, MRSA; that number, it turns out, was larger than the yearly deaths caused by AIDS. It was shocking: an obscure pathogen, unknown to most people, was killing more than the well-known HIV. During 2008, most US newspapers carried some information on antibiotic resistance at least once a week; *Dr. House* had a patient infected by MRSA, and a crucial witness in *Law & Order* died because of an MRSA infection before testifying in court. But the

MRSA crisis more or less subsided, only to be replaced by CRE, carbapenem-resistant enterobacteria. Antibiotic resistance is again trendy partially thanks to CRE.

Antibiotics are drugs and, as such, most people believe that they are mostly used to treat ill people and other animals. Exposure of bacteria to such compounds should, following this logic, occur only in clinical settings. Resistance hinders the ability of antibiotics to effectively cure infections; other than that, in most people's thoughts, resistance has little or no impact. Therefore, resistance is defined not based on the biological change enabling bacteria to survive and even thrive in antibiotic concentrations previously lethal to them; but only in terms of whether or not it can be related to therapeutic failure when using such antibiotic. All the notions above are plagued with misconceptions that had seriously limited research on antibiotic resistance outside the clinical settings. Why would sub-clinical concentrations of antibiotics, and low-level resistance, be of any relevance? Why look for antibiotic resistance in the environment? Who cares if an obscure bacteria from a lake, or from the gut of a wild animal, is resistant to an antibiotic? It is the purpose of this book to answer those questions and dispel those misconceptions. The field of antibiotic resistance in the environment is, however, hardly new: a paper dated more than 40 years ago, warned about the dangers of having resistant coliforms in water supplies, followed by descriptions of all sorts of environmental bacteria resistant to the drugs of that time (*e.g.*, chloramphenicol, tetracycline, ampicillin, nalidixic acid). Looking at more recent papers, the only things that have changed are the resistance figures (always higher) and the names of antibiotics (always stranger). It feels like André Gide was right: "Everything has been said before. But since nobody listens we have to keep going back and begin all over again". However, there has been an interesting twist recently. The arrival of powerful molecular technologies enabled us to look for resistance genes without culturing bacteria (which is important, as less than 1% of soil bacteria have been cultured), and to detect minute quantities of such genes. Using these technologies, a swarm of papers reported resistance genes of all kinds in samples from soil, water, feces ... even very old permafrost. These papers gave the right notion that resistance is everywhere and is ancient; but this notion became distorted as to signify that it is therefore not worrisome to find resistance to, for instance, a new, synthetic drug, in an enteric bacteria from wildlife or a lake sediment. Something very akin to stating that climate has always changed – therefore human influence is irrelevant, an argument so popular amongst those that do not understand the difference between climate and weather. Interestingly, many of those papers emerged when people started to worry about resistance, alarmed also by the lack of new antibiotics – and by the lack of interest from pharmaceutical companies to develop them; only to give room for the grim reports mentioned at the beginning of this preface – including a forecast for a dramatic rise in yearly mortality due to resistance, from 700,000 to 10 million by 2050, along with a 100-trillion US dollars GDP loss, now that everybody seems convinced of the need for financial "incentives" for pharma companies to engage in antibiotic R&D. Curious timing.

Anyway, this book will ride the new wave of interest on antibiotic resistance. A (disproportionately long) introductory first chapter will establish working definitions of antibiotics, resistance and environment, as well as briefly describe the known mechanisms underlying resistance and its spread, and the methods used to investigate the presence of antibiotics and antibiotic resistance in the environment. From there, it will review available evidence of the causes and magnitude of the problem; and why it is a very serious problem indeed. I would try to do so in a way that is accessible for most readers, requiring only basic notions of each of the involved topics. As the subject of antibiotics and environment has repercussions on human and animal medicine, microbiology, ecology, public health, environmental protection, pharmaceutical discovery and policy making, to mention a few, this book will try to provide a basic background for everybody to understand its content. Obviously, to achieve this goal, some sections will be boring to some readers, and some oversimplifications would have to be done. However, I hope that, at the end, all readers, regardless of their background, would get the "big picture" behind the issue of antibiotics and antibiotic resistance in the environment. This is, of course, a big task; and people do not undertake big tasks if not affected by a bit of arrogance (*i.e.*, "an exaggerated sense of one's own importance or abilities"). So, I confess: I am writing this book believing that a training in pharmacology, microbiology and molecular biology, a long time dealing with the clinical side of bacterial resistance, and some recent incursions in the field of resistance in the environment; along with a lifetime in a – so-called – "developing" country, do provide me with enough perspective and insight to convey an integrated view of this problem. Now, let's see if I can deliver.

Mexico City, September 2015

Chapter 1

Definitions and basic concepts

When thinking of "antibiotics in the environment", perhaps a first image that comes to mind is a clandestine dumping of antibiotics from a drug factory into a river in a non-developed country; and when thinking of "antibiotic resistance in the environment", the natural consequence would be to think of an aquatic bacteria under the selection of the dumped antibiotic becoming resistant to the drug, and then causing an outbreak in a neighboring town. The actual scenario is far more complicated and complex: complicated, as there are much more elements at play; complex, as the interactions of those elements are many and multi-directional. Even from the semantic point of view, there could be confusion as to what we call "antibiotic", "resistance" or "environment", so it is important to begin with some working definitions. Although some definitions are not unanimously agreed upon, it is crucial to frame the content of this book within those definitions, in order to avoid misinterpretation.

In addition to formal definitions, this first section will deal with some concepts that are relevant and necessary to understand the reach and limitations of our current knowledge of the topic at hand. What is the "role" of natural antibiotics in microbial ecology? How is the definition of resistance limiting the perspective view of its emergence and evolution? What are the advantages and disadvantages of using molecular-based or culture-based techniques for assessing resistance in the environment? Many of these issues are clearly controversial, and the author's bias will become clear; but by actually stating a position, it is hoped that the readers will be able to reach their own conclusions much more easily.

1.1 ANTIBIOTICS: ORIGINS AND ACTIVITY

"Antibiotic", according to the Merriam-Webster dictionary, is "a substance produced by or a semisynthetic substance derived from a microorganism and able in dilute solution to inhibit or kill another microorganism". This definition would encompass things like natural penicillin (a product of a mold) and ampicillin (a semisynthetic derivative of penicillin); exclude entirely synthetic agents such as sulfonamides and quinolones; and leave in a limbo drugs like chloramphenicol which, although initially discovered as a product of soil bacteria, it is now produced entirely by chemical synthesis. A wider definition from Wikipedia states that "antibiotics [...] are a type of antimicrobial used in the treatment or prevention of bacterial infection", whereas "antimicrobial" is simply "an agent that kills microorganisms or inhibits their growth", which would also include antiseptics and disinfectants. Then there are "antibacterials", "anti-infective chemotherapy", and so on. For the purposes of this book, as there is no evident advantage in discriminating at every sentence between natural and synthetic compounds, an antibiotic would be a chemical agent with a selective toxicity profile, capable of killing or inhibiting the growth of bacteria but mostly incapable of exerting toxicity upon eukaryotic cells at the same concentration (the "magic bullet" imagined by Paul Ehrlich), that is commonly used to treat or prevent bacterial infections. This definition would therefore include all drugs, of natural or synthetic origin, used against bacteria; and would exclude compounds used against viruses, fungi, protozoans or other microorganisms, as well as non-selective biocides, such as disinfectants and antiseptics.

1.1.1 Origin and mechanism of action of main antibiotic classes

Although it is not within the purview of this book to enlist and review the origin and mechanism of action of each class of antibiotics, having an overview included could be helpful for the reader not well versed into this mainly pharmacological area. It may be important to point out that a sort of unifying mechanism of action of bactericidal antibiotics, through a common pathway of generating reactive oxygen species, recently proposed (Kohanski et al., 2007), was first shown to be inconsistent with physiological evidence (Mahoney and Silhavy, 2013), and then most likely to be based on a laboratory artifact (Renggli et al., 2013). It is also important to emphasize that these are the mechanisms of bacteriostatic or bactericidal effects of high, clinically-attainable concentrations of antibiotics; as will be discussed below, this could very well be a human-made situation, with natural antibiotics actually exerting other physiological roles at much lower concentrations. The following paragraphs enlist some relevant information about each antibiotic class, with those that include mostly natural products first. For additional information on the chemistry, pharmacology and clinical uses of each drug, two comprehensive texts can be useful: Bryskier A. (ed.) *Antimicrobial agents, antibacterials and antifungals*; ASM Press, Washington DC, 2005; and Grayson M.L. *et al.* (eds.) *Kucers' The use of antibiotics*, 6th ed; Hodder Arnold, London, 2010.

- *Beta-lactams.* This class includes natural and semi-synthetic penicillins, natural and semi-synthetic cephalosporins and cephamycins (occasionally subgrouped as cephems), carbapenems, monobactams, and beta-lactamase inhibitors. Many members of this class are derivatives of natural products of fungi, *Penicillium* spp. and *Acremonium* (formerly *Cephalosporium*) spp.; while others (cephems, carbapenems, monobactams, beta-lactamase inhibitors) derive from soil bacteria from the genus *Streptomyces* and *Chromobacterium*. However, there is evidence that the genes necessary for the production of beta-lactams by fungi, actually originated from bacteria, making in the end all of these drugs of bacterial origin. With the partial exception of beta-lactamase inhibitors, beta-lactams inhibit the action of peptidoglycan transpeptidases, collectively known as penicillin-binding proteins, or PBPs. As a result, the synthesis of the main component of the bacterial cell wall is halted, while its hydrolysis during bacterial replication, and cellular growth, are not; the osmotic uptake of water occurs without the volume restriction imposed by the cell wall, leading to cytolysis. Apart, beta-lactamase inhibitors are used in conjunction with a penicillin or cephalosporin, so that they protect the actual bactericidal agent from the action of bacterial enzymes responsible for resistance. However, one of these inhibitors, sulbactam, exerts by itself the inhibition of wall synthesis upon the pathogen *Acinetobacter*.

– *Aminoglycosides*. This class of antibiotics includes the natural and semi-synthetic products of soil bacterial of the genus *Streptomyces* and *Micromonospora*. The group's first member was streptomycin, which did open the door for looking into soil bacteria for new antibiotics. This further search led to natural aminoglycosides tobramycin, kanamycin, neomycin, sisomicin, and gentamicin; and semi-synthetic ones, such as amikacin, netilmicin and isepamicin (the "-mycin" suffix indicates a *Streptomyces* product, while the "-micin" one is used for *Micromonospora*-derived compounds). Aminocyclitol antibiotic spectinomycin, a natural product of *S. spectabilis*, is often included in the same group, although its chemical structure is different, as are some details of its mechanism of action. Aminoglycosides bind, sometimes irreversibly, to the 30S ribosomal subunit, leading to inaccurate translation (misreading), impaired proof-reading and/or premature termination of protein synthesis. Aminoglycosides are actively uptaken by components of the bacterial respiratory chain, hence they do not reach inhibitory concentrations intracellularly in anaerobes, or in facultative anaerobes growing under anaerobic conditions.

– *Macrolides*. They include erythromycin, a natural product of the actinomycete *Saccharopolyspora erythraea*; and semi-synthetic derivatives, sometimes classified, for marketing purposes, under individual "classes", such as the "azalide" azithromycin, or the "ketolide" telithromycin. Macrolides reversibly bind the 50S subunit of the bacterial ribosome, specifically the nascent peptide tunnel in the vicinity of the peptidyl transferase center, stalling the ribosome, hence blocking translation. Although chemically very different, lincosamides and streptogramins bind to the same ribosomal site.

– *Lincosamides*. A small class that includes lincomycin, a product of *Streptomyces lincolnensis*; and clindamycin, a semi-synthetic derivative of lincomycin; the main difference – and advantage of clindamycin, is its activity upon anaerobic bacteria. Their mechanism of action is similar to the one of macrolides.

– *Streptogramins*. There are two main subclasses of streptogramins, A and B. Both are products of *Streptomyces* bacteria and, although chemically different, they act in the same way and often synergistically. The combination of quinupristin (streptogramin B) and dalfopristin (streptogramin A) was used in human medicine, while virginiamycin is used in the industrial production of fuel ethanol, and as a "growth promoter" food additive for livestock. Their mechanism of action is similar to the one of macrolides.

– *Amphenicols*. Chloramphenicol, a product of *Streptomyces venezuelae*, is the main representative of this group; synthetic derivatives (chloramphenicol used today is chemically synthesized itself) include florfenicol, used only for veterinary purposes; and thiamphenicol, used for humans in some countries, and for animals in others. Amphenicols bind to the 23S rRNA of the 50S ribosomal subunit, inhibiting the peptidyl transferase activity of the bacterial ribosome.

– *Tetracyclines*. Natural (chlortetracycline, from *Streptomyces aureofaciens*; oxytetracycline, from *S. rimosus*) and semi-synthetic (minocycline, tigecycline, the later considered a "glycylcycline") are members of this group. Tetracyclines inhibit bacterial synthesis of proteins by binding to the small ribosomal subunit, blocking the attachment of aminoacyl-tRNA to the A site; tetracyclines bind to both, the 30S bacterial subunit, and the 40S eukaryotic one, but bacteria uptake tetracyclines actively, leading to much higher intracellular concentrations.

– *Glycopeptides*. Vancomycin, a natural product of soil bacterium *Amycolatopsis orientalis*, was the first member of this group, followed by other natural (teicoplanin, ramoplanin, from *Actinoplanes* spp.; avoparcin, from *Streptomyces candidus*) and semi-synthetic (*e.g.*, telavancin, oritavancin), products. Glycopeptides inhibit the synthesis of the cell wall of gram-positive bacteria; gram-negatives are usually non susceptible due to the inability of glycopeptides to cross the outer membrane. These antibiotics bind to the D-alanyl-D-alanine moieties at the end of the short peptide hanging from acetylmuramic acid, before the cross-linking of peptidoglycan; the attached antibiotic prevent the cross-linking itself.

– *Lipopeptides*. Polymyxins (B, and E, known as colistin), products of *Paenibacillus polymyxa*; and daptomycin, obtained from *Streptomyces roseosporus*, are included in this group. Lipopeptides seem to alter the architecture of the phospholipid bilayer of the cell membrane; while polymyxins first attach to the lipopolysaccharide in the outer membrane of a few gram-negatives, and then gain access to the cell membrane; daptomycin binds to the cell membrane of gram-positives in a calcium-dependent manner. Lipopeptides are often regarded as "last-option" antibiotics: polymyxins are used only against multi-resistant bacteria (*i.e.*, carbapenem-resistant Enterobacteriaceae, *Pseudomonas aeruginosa* and *Acinetobacter* spp.), and daptomycin against methicillin-resistant *Staphylococcus aureus* (MRSA) and vancomycin-resistant enterococci (VRE).

– *Fosfomycin*. Fosfomycin is a small molecule isolated from *Streptomyces fradiae*; it inhibits the synthesis of bacterial cell wall, through the inhibition of MurA (UDP-acetylglucosamine-3-enolpyruvyltransferase), acting as a phosphoenolpyruvate analog. Fosfomycin is not widely used, despite a wide spectrum, low toxicity and low resistance rates. In countries where it is a preferred option (such as Spain, as the antibiotic was discovered there), it was mostly used against urinary tract infections; however, fosfomycin has shown relevant activity against multi-resistant organisms that are common in hospital settings, and is now regaining attention as an option in the management of infections caused by such bacteria.

– *Rifamycins*. Rifampicin (or rifampin), rifabutin, rifapentine, and orally-unabsorbable rifaximin, are all derivatives of rifamycin, a natural product of *Amycolatopsis rifamycinica* (formerly *A. mediterranei*, formerly *Nocardia mediterranei*, formerly *Streptomyces mediterranei*). They selectively inhibit bacterial RNA-polymerase. Rifamycins have mostly been used against tuberculosis, but have also been used against multi-resistant staphylococci and pneumococci. Rifaximin is used against intestinal bacteria; it should be used only for intestinal "sterilization" prior to gut surgery, and the management of hepatic encephalopathy, although it is also abused as an anti-diarrheic agent.

– *Sulfonamides (DHPS inhibitors)*. Prontosil, a prodrug of sulfanilamide, was the first synthetic antibiotic, patented in 1932. While "sulfonamide" refers to a chemical functional group, common to a very wide variety of molecules, many of clinical relevance (*e.g.*, diuretics, sulfonylureas, antiretrovirals, anti-inflammatory drugs), the term sulfonamide is often used to refer only to those of antimicrobial properties. Sulfonamides are structural analogs of *p*-aminobenzoic acid (PABA), hence acting as competitive inhibitors of dihydropteroate synthase (DHPS), an enzyme that forms dihydropteroate (a precursor of folic acid) from dihydropterid-inmethyl phosphate and PABA. There are dozens of sulfonamides, varying mostly in pharmacokinetic properties; today, sulfamethoxazole is the most widely used (in association with trimethoprim; see below); silver sulfadiazine is also used topically. DHPS is not present in mammals, as we need to have dietary folates.

– *Trimethoprim (DHFR inhibitors).* At the end of the pathway that joins PABA and dihydropteroate diphosphate to form dihydropteroic acid, enzyme dihydrofolate reductase (DHFR) sequentially converts folic acid into dihydrofolate and then tetrahydrofolate, the actual cofactor for the synthesis of purines, thymidylic acid, and several amino acids. DHFR is inhibited by trimethoprim, the most widely used drug of this class, that also includes much less used brodimoprim, and failed drug iclaprim. Trimethoprim is often used in combination with the sulfonamide sulfamethoxazole; the combination, known as co-trimoxazole, was supposed to be synergic and less prone to select for resistance, which end up to be false expectations. While eukaryotes have also DHFR, trimethoprim and related drugs inhibit prokaryotic DHFR at concentrations several orders of magnitude lower.

– *Nitrofurans.* This group of synthetic molecules is mostly represented by nitro-furantoin, a drug only used against lower urinary tract infections; interest in nitrofurantoin has increased recently, as resistance among uropathogenic *E. coli* remains very low (<10%), despite decades of clinical use. The mechanism of action of nitrofurans seem to involve the formation of a reduced, reactive intermediate, that in turn disrupts DNA, ribosomes and respiratory chain. The formation of this intermediate depends on bacterial reductases, hence it is mainly produced inside bacterial cells. Other drugs in this group are nitrofurazone, used topically; and furazolidone, used against enteric pathogens.

– *Quinolones.* The quinolone antibiotic class is one of the latest ones to be introduced into clinical use; paradoxically, the first one of the family, nalidixic acid, is not quite a quinolone, but a naphthyridone. Second-generation quinolones are also the first fluoroquinolones, with an added fluoride atom: norfloxacin, ciprofloxacin, ofloxacin (and its homochiral formulation levofloxacin) and enrofloxacin, the latter used only on animals. These agents act as inhibitors of class 2 topoisomerases, gyrase and topoisomerase IV, although their bactericidal action is mostly due to gyrase inhibition. It is important to understand how class 2 topoisomerases work: they bind to DNA, cleave both strands (hence "class 2"), make other, intact double-stranded DNA pass across the cleavage, and bound the cleaved DNA back. While in the presence of quinolones, topoisomerases do cleave DNA, but are unable to rebound it; quinolones induce double-strand breaks in DNA, that are very difficult to repair. Many fluoroquinolones were introduced into clinical use, only to be withdrawn some few months later due to adverse effects. Nalidixic acid and second-generation quinolones are particularly active against gram-negative bacteria, although levofloxacin has been inadequately used against gram-positives. A third generation of fluoroquinolones (moxifloxacin, gemifloxacin) are better inhibitors of the gyrase of gram-positives, and are mostly used against respiratory tract diseases.

– *Oxazolidinones.* The first drug of this group, linezolid, was approved for clinical use in 2000, making oxazolidinones the latest antibiotic class to be introduced (with a nearly 40-year gap between them and the first quinolone). Along with tedizolid, the only other oxazolidinone approved up to this date, these molecules block the initiation of protein synthesis, by binding to the 23S portion of the 50S ribosomal subunit, close to the binding site of amphenicols. Oxazolidinones are active only against gram-positive bacteria. Other drugs in this group that are likely to reach the market are posizolid and radezolid.

– *Nitroimidazoles.* Although mostly used against protozoans, metronidazole is also active against anaerobic bacteria (and facultative anaerobes under anaerobic conditions). A reduced derivative, only produced in the absence of free oxygen, interacts with DNA resulting in cell death. An exception occurs in *Helicobacter pylori*, against which metronidazole is also effective, but that thrives in environments still too rich in oxygen; in this species, an oxygen-insensitive nitroreductase converts metronidazole into its reduced toxic derivative. Related drug tinidazole supposedly acts in the very same way, although is mainly used against protozoal infections.

Aside from the antibacterial properties of antibiotics, briefly described above, some of these drugs have been used for an entirely different purpose: by reasons that are not fully understood, sub-therapeutic doses of some antibiotics can promote the growth of farm animals. Possible ways antibiotics can achieve this include: (a) reducing microbial use of nutrients, (b) diminishing the thickness of the intestinal wall, thus enhancing uptake of nutrients, (c) preventing some sub-clinical infections, and (d) reducing microbial metabolites that can depress animal growth (Laxminarayan et al., 2015). Although the actual mechanism of animal growth promotion is not clear, this represents by far the main use of antibiotics, as will be analyzed in Chapter 3.

1.1.2 Antibiotics: chemical warfare, intercellular signaling, prebiotic remains?

The term "antibiotic", introduced by streptomycin "discoverer" Waksman, could be understood as derivative of *antibiosis*: "antagonistic association between organisms to the detriment of one of them or between one organism and a metabolic product of another". This suggests, as it is a common belief, that natural antibiotics are a sort of chemical weapon used by producing bacteria, to colonize new environmental niches and/or to keep invaders at bay from an already colonized place. In addition to its simplicity, this explanation has the appeal of assuming that chemical warfare is something natural and not an abomination. Furthermore, in such war-like scenario, resistance seems like a normal, ubiquitous trait to emerge wherever an antibiotic-producing, aggressive bacterial strain is around. However, natural antibiotics are much more complicated than that.

Julian Davies was the first to suggest that antibiotics are unlikely weapons for chemical warfare. Based on the complexity of their synthetic pathways, the fact that they are mostly produced when bacteria are in stationary phase, and the effector activity many of them have upon gene expression and transfer, Davies proposed that antibiotics could be remnants of prebiotic molecules that interacted with early nucleic acids, even before the emergence of ribosomes, and are now adopted as secondary regulators (Davies, 1990). Antibiotics act as bacteriostatic or bactericidal agents only at concentrations much higher than those found in nature. At sub-inhibitory concentrations, antibiotics are known to act as signaling molecules, having very subtle effects on gene expression at different levels. Among the many aspects of bacterial physiology that are affected by low concentrations of antibiotics, are those related to quorum, biofilm formation, and others involved in the coexistence of different bacterial species in the same ecological niche (Sengupta et al., 2013). At the clinical perspective, subinhibitory concentrations of antibiotics act upon the expression of virulence determinants (Linares et al., 2006). As the notion of coexistence and non-competitiveness clash with the anthropocentric warfare view, antibiotics are still mainly regarded as microbial chemical weapons, despite the evidence in contrary. To be fair, there are some few known examples where antibiotics – none used clinically – seem to be produced to suppress the growth of competitor microorganisms (Sengupta et al., 2013).

While the controversy around the actual role of natural antibiotics may seem purely academic, it holds the key to understand the emergence of resistance and its role in open environments. Again, it was Julian Davies who first demonstrated the biochemical similarity between some of the resistance mechanisms found in antibiotic-producing bacteria and in clinically-relevant ones (Benveniste and Davies, 1973). There is now some genetic evidence of the presence of antibiotic-resistance genes in antibiotic-producing bacteria; protection from their own products, as well as having some role in the very biosynthesis of the antibiotics themselves, are among the proposed roles of these genes (Sengupta et al., 2013). However, many resistance genes have been reported in non-producing, environmental bacteria, the so-called "resistome". If natural antibiotics occur only at very low concentrations, it is unlikely that they exert a selective pressure for a full-resistant phenotype. Are antibiotics actually present at high concentrations in nature? Are there other, non-antibiotic agents that select for antibiotic-resistance genes? Or could it simply be that we are using the word "resistance" in a very loose way? All this will be discussed in the next chapter.

1.2 RESISTANCE: WHAT IT IS AND HOW WE MEASURE IT

As the main – recognized-role of human-made antibiotics is to cure infections, resistance has always been considered from a clinical, instead of a biological point of view. This is to say that it is not merely a matter of comparing the ability of one strain against another, to withstand an antibiotic at a given concentration; but to try to correlate such ability to the likelihood of the antibiotic to fail if used therapeutically against the "resistant" strain. This notion has resulted in dangerous generalizations and oversimplifications. For instance, a set of concentration breakpoints have been established: if a strain grows *in vitro* in the presence of antibiotic concentrations above such breakpoint, it is deemed resistant. Such breakpoints purportedly consider the antibiotic concentrations reached clinically when the drug is administered at standard doses; also, the cure rate when using the drug, related to the inhibitory concentrations for each isolate in a number of patients. However, for these breakpoints to be adequate, they should consider the wide variation of tissular concentrations achieved by an antibiotic within a single patient; and the even wider variation between different patients. Also, the pharmacokinetics and pharmacodynamics of each drug should be considered, along with the dosing schemes. If all these issues were to be included, the breakpoint list would be endless and completely unpractical; therefore, all we have are single breakpoints for each antibiotic against each major bacterial group of clinical relevance. For the subject of this book, that is, bacterial resistance in the environment, this poses great difficulties, as for many environmental, innocuous bacteria, there are simply no established breakpoints, in addition of the whole concept of resistance, from the clinical point of view, would be mostly irrelevant. Moreover, the limitations do not end there.

The activity of antibiotics *in vitro* is generally assessed in two ways: by determining the minimum inhibitory concentration (MIC) of the antibiotic, usually in liquid media; or by measuring an inhibitory halo produced by the antibiotic diffusing from a small paper disk into a solid culture media. Media used for these assays have been selected to optimize bacterial growth, and to minimize their interference with antibiotic activity (*e.g.*, devoid of para-aminobenzoic acid, as it antagonizes the effect of sulfonamides). Pure cultures, at standardized innocula, are tested; bacteria are therefore growing in artificial conditions entirely different than they do in clinical or environmental settings. When assessing the MIC, bacteria are suddenly exposed to high antibiotic concentrations, reducing the ability of adaptive responses to be activated; and in a way that is very different from the gradual exposure that typically occurs in nature (the disk diffusion technique allows for a gradual exposure, as the antibiotic diffuses slowly from the disk into the agar medium). The effect of the antibiotic is measured in the very short term −18 to 24 hours, but as short as 4 hours if using an automated system. This would hinder the ability to detect slow-growing varieties, as well as hetero-resistance (see below). Bacterial growth is measured just by the turbidity they produce in liquid media, by eye or using a nephelometer; or by trying to assess the diameter of an often irregular, diffuse halo surrounding a paper disk. Finally, MIC determinations rely on series of double-fold dilutions, that analyze with detail the effect of low concentrations (*e.g.*, 0.01, 0.02 and 0.04 µg/mL), but that leave great gaps at high concentrations (*e.g.*, 32, 64, 128 µg/mL). In any case, we end up with a MIC value, or a halo's diameter. These values are then to be compared to breakpoints' tables that enlist antibiotics and bacterial groups (*e.g.*, enterobacteria, non-fermentative bacteria, staphylococci), so that we can interpret whether a 12 µg/mL MIC, or a 17-mm halo, is indicative of susceptibility of resistance of that particular species and for that particular antibiotic. These tables change from time to time, resulting in the curious paradox of having a phenotype classified as resistance one year, and as susceptible the following year (this has had a very negative impact in assessing the evolution of resistance along time, especially for rapidly changing breakpoints, such as penicillin susceptibility in pneumococci). And, to make it all worst, each geopolitical region has its own set of breakpoints: the US and, by extension, most of the American countries (America, by the way, is the name of a continent, not of a country), follow the ones set by the Clinical Laboratory Standards Institute (CLSI); the EU have their owns (EUCAST); as do the UK (BSAC); and they do not always match. As a result, a bacterial strain can be considered resistant in Europe, but susceptible in America.

For an adequate assessment of resistance in the environment, a biological rather than clinical definition should be used. A useful approach is to consider the natural variation of antibiotic activity upon a large number of isolates of a given bacterial species. This would enable the separation of susceptible and resistant bacteria within each species or other relevant taxa, independently of the clinical nuances of achievable plasma concentrations and relatedness to treatment outcomes. An ECOFF, or epidemiological cut-off value for resistance breakpoint has been proposed (Martínez et al., 2015), aimed at a biological rather than clinical description of resistance. However, as this data is mostly missing, resistance would have to be defined through this book, based on the clinical breakpoints, when available. When referring to diminished susceptibility not reaching said breakpoints (*e.g.*, the one conferred by plasmid-borne *qnr* quinolone-resistance genes); or to data obtained using selecting concentrations of antibiotics different from said breakpoints (*e.g.*, using agar plates containing ampicillin concentrations of 50–100 μg/mL, while the breakpoint for resistance in enterobacteria is 32 μg/mL); or to species and/or antibiotics for which breakpoints are not available (*e.g.*, streptomycin for *E. coli*), an aclaratory note would be made, along with the reference to "resistance", within quotation marks.

1.2.1 Resistance mechanisms, horizontal gene transfer, and adaptive responses

Bacteria can withstand the effect of antibiotics in many different ways. This section will first review the main biochemical and physiological mechanisms through which bacteria can survive and even thrive in the presence of antibiotics; then, it will review the genetic phenomena that allow such biochemical and physiological mechanisms to arise and spread. Although of considerable interest, this section will not further discuss other related phenotypes, such as tolerance (an increase in the concentration needed to kill bacteria, while the inhibitory concentration remains unchanged); subsistence (the ability to use antibiotics as a carbon or nitrogen source); and dependence (a rare phenomenon where affected bacteria can grow *only* in the presence of an antibiotic). Variations of these phenomena, and even of the ones that will be described in following paragraphs, can be considered as "noninherited" resistance mechanisms (Levin, 2004), which are of great interest but that are still far from being adequately understood.

1.2.1.1 Three main kinds of resistance: intrinsic, acquired, and adaptive

There are many ways in which resistance mechanisms can be classified. For the purposes of this book, resistance mechanisms will be categorized as intrinsic, acquired, or adaptive. It is important, however, to state that the boundaries for each of these three categories can be diffuse and some times confusing. This confusion is particularly relevant to the subject of this book, as it seems to pervade the research in the area, as will be further discussed.

1.2.1.1.1 Intrinsic resistance

Intrinsic resistance can be defined in different ways. It can be thought as an inherent characteristic of a given bacterial species, to be unaffected by an antibiotic at concentrations achieved clinically. Many cases of intrinsic resistance are related to permeability issues: the outer membrane of *Pseudomonas aeruginosa*, for instance, is mostly impermeable to aminopenicillins (ampicillin or amoxicillin), as is the outer membrane of many gram-negative bacteria to macrolides and glycopeptides. But there are some other mechanisms underlying intrinsic resistance: the inability of anaerobic bacteria to uptake aminoglycosides render them intrinsically resistant to such compounds; a

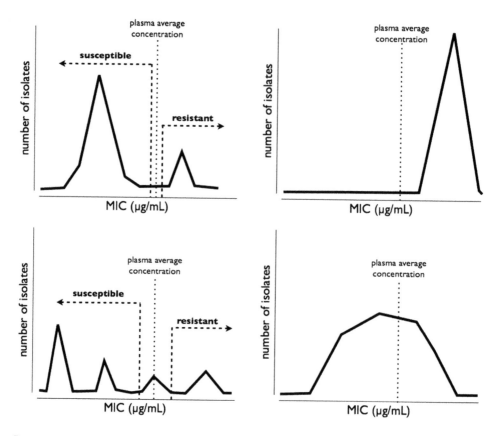

Figure 1.1 **Definitions of resistance; a graphic summary.** Top left, the easiest scenario, where members of a bacterial species have either, a very low or a very high MIC, and plasma (or tissue) concentrations are always enough to inhibit the low-MIC varieties. A breakpoint for resistance follows both, the biological variation, and the clinically-achievable concentrations. Top right, a clear example of intrinsic resistance, where all members of a bacterial species are inhibited by an antibiotic, but at concentrations too high to be reached clinically. Bottom left, the most usual scenario with clinical isolates of a bacterial species, having a variety of phenotypes, some under, some above, and some even overlapping clinical concentrations. This variety some times creates the messy definition of "intermediate susceptibility", a MIC range between resistant and susceptible, that physicians seldom know how to use. From the biological point of view, all three peaks at the right have gained resistance; but from the clinical point of view, only the one at the far right is truly resistant. Bottom right, a very wide distribution of MICs that makes the determination of resistance and susceptibility a very hard – and often useless task.

complex and distinctly different set of enzymes involved in peptidoglycan synthesis in enterococci render this genus intrinsically resistant to cephalosporins (Vesić and Kristich, 2012). When considering a timeframe, intrinsic resistance is a defining characteristic of both, a bacterial species and an antibiotic, a characteristic that has not changed in time, especially within the "antibiotic era". Intrinsic resistance defines the original spectrum of activity of each antibiotic; there is no known antibiotic capable of inhibiting all bacterial species at clinically relevant concentrations. For the purposes of this book, intrinsic resistance is considered to be irrelevant, as it is an unchanged, inherent feature of each bacteria/antibiotic combination. The mechanisms mediating intrinsic resistance are part of the defining characteristics of each bacterial species, residing in housekeeping genes, extremely unlikely to be transferred from an intrinsically resistant bacteria to a susceptible one. Intrinsic resistance cannot be considered a public health threat, as it is an inherent bacterial characteristic – or an inherent antibiotic limitation.

While the examples provided in the last paragraph are straightforward enough, some other bacterial abilities may qualify as "intrinsic" resistance, but will be considered separately in this section. Three examples are singled out below: (1) increased unspecific efflux, elicited by the presence of an antibiotic; (2) the persistence of biofilms; and (3) the almost universal presence of chromosomal beta-lactamases in some enteric bacteria. The first two will be discussed under the adaptive resistance category; the third one within the acquired mechanisms.

1.2.1.1.2 Acquired resistance

The accidental discovery of penicillin occurred because an agar plate with *Staphylococcus aureus* growing all across, became contaminated with a *Penicillium* mold growing in an edge of the plate. At that time, most *S. aureus* isolates had a penicillin MIC $\ll 1\,\mu g/mL$. Strains reported as resistant to penicillin, in 1942 and from that date forward, have penicillin MIC $>100\,\mu g/mL$. These strains have acquired a resistance phenotype, some times even during the course of an antibiotic treatment, within a single patient. This is of course named acquired resistance. In addition of being a recent acquisition, these traits are from that moment on, more or less stably inherited to daughter cells; and usually qualify as full-resistance (*i.e.*, well above clinical resistance breakpoints). Genetically, these traits can be acquired through two main mechanisms that will be further discussed: mutations and horizontal gene transfer; this section will deal with the biochemical mechanisms that enable bacteria to resist the effects of antibiotics.

While several hundreds of resistance genes have been characterized up to this date (and probably a few more while this book is being prepared), all of them can be mechanistically categorized in four groups: (1) enzymatic inactivation of the antibiotic, mostly of naturally-occurring antibiotics; (2) protection or modification of the target of antibiotic action; (3) diminished accumulation of the antibiotic achieved by active efflux and/or diminished permeability; and (4) acquisition of a by-pass route for an antibiotic-blocked pathway, or overproduction of enzymes within such pathway. There are possibly hundreds of reviews on this matter, from one of the earliest but still useful (Foster, 1983); to a recent extraordinarily comprehensive listing of acquired resistance genes (van Hoek et al., 2011). The following is just a brief overview on the subject, with most information coming from the van Hoek paper, except when specifically stated.

– *Resistance to beta-lactams.* Resistant bacteria can (1) enzymatically inactivate beta-lactams, using hydrolases known as beta-lactamases; or (2) produce a modified PBP (paradoxically, a "penicillin-binding protein" that does not bind to penicillin) that enables the synthesis of peptidoglycan even in the presence of beta-lactams. Additionally, diminished accumulation can yield a low-level "resistance" or, if coexisting with a beta-lactamase gene, enhance the protective spectrum of the enzyme. There are around 1,000 known beta-lactamase genes, that encode, from very narrow spectrum enzymes, capable of hydrolysing only penicillins (also known as penicillinases); to extended-spectrum beta-lactamases (ESBL), capable of inactivating third-generation cephalosporins; to carbapenemases, that inactivate all beta-lactams, including carbapenems. Some of these enzymes have a zinc ion within their active site (metallo-beta-lactamases); some are resistant to beta-lactamase inhibitor clavulanate. Most are plasmid-borne, but *ampC* genes are chromosomal in many enterobacteria and in *Pseudomonas aeruginosa* (although also present in plasmids and somehow linked to other mobile genetic elements (Jacoby, 2009)); AmpC enzymes can inactivate several cephalosporins, and are often inducible, even by clavulanate, resulting in a wide-spectrum, clavulanate-resistance phenotype. Mutations in the regulatory genes result in the constitutive overproduction of AmpC, along with increased resistance. As to altered PBPs, these are commonly found in two clinically relevant bacterial groups: (a) streptococci, particularly *S. pneumoniae*, which altered PBP genes were mobilized by transformation, and a mosaicism phenomena resulting from homologous recombination; also, low-affinity PBPs have been found to be plasmid-encoded in enterococci (Raze et al., 1998); and (b) *S. aureus*, mediating the phenotype known as methicillin-resistance (an archaism, as methicillin is no longer in clinical use; "methicillin-resistance" therefore refers to resistance to anti-staphylococcal penicillins, such as oxacillin, and actually includes most beta-lactams), the well-known MRSA.

- *Resistance to aminoglycosides.* Most acquired resistance to aminoglycosides depend on the enzymatic inactivation of the antibiotic, through acetyl-, phosphoryl- or nucleotidyl-transferases; altogether, there are more than 150 different known genes encoding such enzymes. A few methyltransferases, encoded by *rmt* genes, have been more recently described in gram-negatives; and a bifunctional acetyl- and phosphoryl-transferase, that inactivates most aminoglycosides in clinical use, is common in gram-positive cocci.

- *M and MLS$_B$ resistance.* In the MLS$_B$ (macrolide, lincosamide, streptogramin B) antibiotic group, macrolides are the most diverse and commonly used ones. Macrolide resistance is mediated by two main mechanisms: specific efflux, mediated by *mef* genes, whose products expel only macrolides (M phenotype); and ribosomal protection via methylation of the 23S rRNA by methylases encoded by *erm* genes, which result in cross-resistance towards lincosamides and streptogramins (MLS$_B$ phenotype), as these antibiotics bind to the same ribosomal region. From the clinical point of view, *mef*-mediated resistance is of low level and, perhaps, of little relevance, as increased MIC are still below plasmatic and tissular concentrations (Anzueto and Norris, 2004). However, as many *erm* genes are inducible, resistance breakpoints in the clinical lab have been set low enough to include fully-resistant, *erm*-bearing bacteria along with *mef*-bearing "resistant" organisms. Aside, a short list of macrolide-inactivating enzymes have been recently described, although many of them in Enterobacteriaceae and other gram-negatives considered to be intrinsically resistant to these antibiotics (except for *vat* transferases, found in gram-positive cocci).

- *Resistance to amphenicols.* The most common mechanism of resistance to chloramphenicol and thiamphenicol is enzymatic inactivation through acetyltransferases, CAT; florfenicol, only used in animals, is not affected by these enzymes. A few other genes encode for efflux systems: *cmlA*, also ineffective against florfenicol; and *floR*, which confers resistance to chloramphenicol and florfenicol alike. A ribosomal methyltransferase encoded by *cfr* genes, whose action prevents the binding of florfenicol to the ribosome, is important as it also mediates resistance to other clinically-relevant antibiotics, such as linezolid.

– *Tetracycline resistance*. There are more than 40 *tet* genes mediating tetracycline resistance. Most of them encode specific efflux pumps, while about 10 mediate a ribosomal protection mechanism, and 5 or so an inactivating enzyme found only in gram-negatives. Although *tet* genes have not been implicated in tigecycline resistance, the extensive use of tetracyclines for agricultural purposes, along with the linking of *tet(M)* and *erm(B)* genes in mobile genetic elements (Moritz and Hergenrother, 2007), have fostered the interest in tetracycline resistance determinants.

– *Resistance to glycopeptides*. Being antibiotics that uniquely bind to a substrate rather than an enzyme or ribosome component, resistance mechanisms are also peculiar. The binding of glycopeptides to the D-Ala-D-Ala terminus of a peptidoglycan precursor is prevented by changing this terminus to D-Ala-D-lactate or, less commonly, to D-Ala-D-Ser. The changing of D-Ala-D-Ala to D-Ala-D-Lac involves at least a D-D dipeptidase to remove the last D-Ala, a lactate dehydrogenase that synthesize the D-lactate, and a ligase that binds the D-lactate. *vanA* and *vanB* are actual operons; although some of the genes included into these came probably from soil bacteria (*Paenibacillus popilliae*), the codon usage of the genes differs, suggesting a different origin for each of them. The expression of these operons is induced by the glycopeptides themselves: *vanA* is induced both by vancomycin and teicoplanin, while *vanB* is only induced by vancomycin (Fraimow, 2003).

– *Resistance to lipopeptides*. Acquired resistance to polymyxins are often mediated by changes in outer-membrane lipopolysaccharides, or even the complete loss of them, which are necessary for the initial binding of the drug to the bacterial envelope (Olaitan et al., 2014). Although such modifications can be accompanied by diminished virulence and resistance to other antibiotics, they still pose a significant health threat. Resistance to daptomycin is still rare; resistant enterococci isolates have mutations in enzymes involved in phospholipid metabolism, as well as in a putative membrane protein (Arias et al., 2011); no inactivation was detected in a previous report, although *Actinoplanes utahensis* is capable of deacylating daptomycin to an inactive derivative (Montero et al., 2008). In *S. aureus*, resistance is also linked to membrane and wall changes: genes responsible for the synthesis of lysyl-phosphotidylglycerol and for the D-alanylation of teichoic acids are among those mutated in resistant isolates (Bayer et al., 2013).

– *Resistance to DHPS inhibitors*. Mutations in the chromosomal dihydropteroate synthase gene, *dhps*, have been identified as causing sulfonamide resistance; these can go from point mutations in *E. coli*, to a 10% difference in nucleotide sequence found in *Neisseria meningitidis*, more likely to have arisen by transformation-recombination. However, in most pathogens is much more common to find genes *sulI* and *sulII*, both closely related but with significant sequence divergence from chromosomal *dhps* genes. These *sul* genes encode drug-insensitive enzymes that by-pass the effect of sulfonamides; they are often found in plasmids and, more precisely, in the conserved regions of integrons (Huovinen et al., 1995).

– *Resistance to DHFR inhibitors*. A number of *dfr* genes, encoding a non-allelic, drug-insensitive dihydrofolate reductase, can by-pass the blockage exerted by trimethoprim and related agents upon the synthesis of tetrahydrofolate. Most *dfr* genes are grouped into families A and B, and are mostly found in gram-negatives; while a few other individual genes (*dfrC, dfrD, dfrG* and *dfrK*) coming from gram-positives.

– *Resistance to nitrofurans.* Being nitrofurans pro-drugs that must be "activated" by bacterial reductases, the loss of such enzymes (nitroreductases encoded by *nfsA* and *nfsB*) is an obvious and easy way to acquire resistance. However, such a loss has a high fitness cost (Sandegren et al., 2008) that makes resistant bacteria very weak, hence their low prevalence. A similar resistance mechanism has been reported for related drug furazolidone (Martínez-Puchol et al., 2015). It is likely that intrinsically resistant bacteria lack this kind of reductases. A 30-years old report (Breeze and Obaseiki-Ebor, 1983) of plasmid-mediated nitrofuran resistance did not advance much on the possible biochemical mechanism; however, plasmids were more recently reported to be linked to nitrofurantoin resistance in clinical isolates of uropathogenic *E. coli* (Arredondo-García and Amábile-Cuevas, 2008).

– *Resistance to quinolones.* Most clinically relevant, quinolone resistant bacteria, bear mutations on target genes, *i.e.*, those encoding affected topoisomerases (*gyr* and *par* genes in enteric bacteria). While single mutations are often enough to confer resistance to nalidixic acid, two or more are necessary to confer resistance to fluoroquinolones. Many combinations of mutations are known to enable the survival of bacteria in previously inhibitory concentrations of the drugs (Fuchs et al., 1996). During the early years of quinolone usage, the lack of plasmid-mediated resistance (Courvalin, 1990) was highlighted as an interesting feature of this class of antibiotics, perhaps predicting a slower spread of resistance among clinically relevant bacteria. The recessive nature of the trait supported the notion that horizontal transfer was not likely. This was shown to be wrong, as the acquisition of mutated topoisomerase genes through transformation was demonstrated among streptococci (Balsalobre et al., 2003, Ferrándiz et al., 2000) (homologous recombination after transformation enabled the replacement of the wild-type gene). Also, a number of plasmid-borne, horizontally transferable *qnr* genes have been reported, starting from the 1990's. These genes encode pentapeptide repeat-containing proteins that possibly prevent the binding of topoisomerases to DNA, preventing DNA cleavage, but without actually inhibiting the enzymes' activity somehow. *qnr* genes do not confer full fluoroquinolone resistance, and only increase the MIC, from $\leq 0.01\,\mu g/mL$, to 0.12–$0.5\,\mu g/mL$ (resistance breakpoints are usually $4\,\mu g/mL$ or higher). However, they seem to play an important role in fostering the ability of bacteria to survive in higher concentrations of quinolones, increasing the likelihood of gaining full resistance by other means. A quinolone-modifying enzyme, encoded by *aac(6′)-Ib-cr*, derived from an aminoglycoside acetyltransferase gene, has also been found in plasmids, mediating a low-level resistance phenotype as well (Strahilevitz et al., 2009).

– *Resistance to oxazolidinones.* Resistance to linezolid is still rare. A number of mutations in the domain V of the 23S rRNA gene, and in L3 and L4 ribosomal proteins, have been found in clinical resistant isolates (Campanile et al., 2013). Most interesting for the purposes of this book, is the ribosomal methyltransferase encoded by the *cfr* gene: this gene is plasmid-borne, and confers resistance to antibiotics used in veterinary medicine, such as florfenicol, lincosamides and

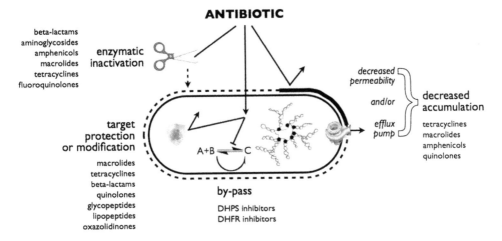

Figure 1.2 **Mechanisms of acquired, specific resistance; a graphic summary.** Four general mechanisms of resistance to individual antibiotics (or individual classes of antibiotics): enzymatic inactivation, decreased accumulation (in turn derived from decreased permeability and/or efflux pumps), target protection or modification, and pathway by-pass. This figure does not include a fifth mechanism, *i.e.*, the loss of target (as in lipopeptide-resistant bacteria that do not produce LPS) or of activating enzyme (as in nitrofuran-resistant bacteria that do not produce the nitroreductases needed for activation of the prodrug). Also, non-specific mechanisms are not included, such as unspecific efflux systems that provide a multi-resistance phenotype, as discussed below.

pleuromutilins. Staphylococcal strains of animal origin may be acting as reservoirs of these genes, that are now found in linezolid-resistant clinical isolates (Tewhey et al., 2014). The activity of tedizolid, a newer oxazolidinone, is not affected by *cfr* genes, and minimally affected by ribosomal mutations. A recently reported *otprA* gene, found in enterococci both, from food-animals and humans, confer resistance to both, linezolid and tedizolid, as well as to amphenicols (Wang et al., 2015).

–   *Resistance to metronidazole*. In anaerobic *Bacteroides*, *nim* genes have been identified as able to confer resistance to metronidazole. Apparently, *nim* genes encode a reductase that reduces the 5-nitroimidazole to a 5-amino inactive derivative (Soares et al., 2012).

1.2.1.1.3 Adaptive resistance: stress responses

Bacteria often change suddenly from one environment to another, each change involving conditions that vary widely. To survive these variations, many bacterial species are equipped with complex, overlapping, but unspecific defense systems; some of them can confer "resistance" to several antibiotics, usually by diminishing the permeability of the outer membrane, and/or by overexpressing efflux pumps. These "resistance" phenotypes are unspecific, providing protection towards a variety of antibiotics and other xenobiotics; are transient, their effects lasting only the duration of the exposure to the inducing stimuli; and increases in antibiotics' MICs are only mild and rarely above full-resistance clinical breakpoints. From the clinical diagnostic point of view, these mechanisms do represent a challenge, as inducing conditions that can be found *in vivo*, are seldom present *in vitro*, being therefore undetectable by the typical susceptibility assays. Mutations in the regulatory genes that result in the constitutive expression of these systems can be construed as "acquired resistance"; however, due to the unspecificity of their protective capabilities, there could be a vast variety of selective and maintenance pressures favoring such mutants, other than antibiotics themselves. Hence, the presence of such mutations can hardly be related to an specific antibiotic, or even to antibiotics as a group. Finally, a wide variety of unrelated mutations can enhance the ability of bacteria to tolerate slightly inhibitorial concentrations of antibiotics; affected genes can be called "susceptibility genes", instead of "resistance" ones, but the end result is the same if mutated: low-level resistance, but allowing for an additive nature that can result in full-resistance (Girgis et al., 2009).

Two well-known regulons of *E. coli* and related enteric bacteria that are involved in antibiotic resistance phenotypes, are the *soxRS* regulon, governing the response to superoxide stress; and the *marRAB* regulon, that regulates a response to a number of chemical insults (Demple and Amábile-Cuevas, 2003). These two regulons overlap extensively. The genes for the efflux system AcrAB-TolC, and *micF*, a gene that encodes an antisense RNA that post-transcriptionally represses the expression of OmpF, are included in both regulons. Overexpression of AcrAB-TolC and repression of OmpF result in decreased accumulation of several antibiotics and, in turn, diminished susceptibility. "Resistance" achieved through these mechanisms is rarely enough to be of clinical relevance alone (although *marR* was formerly called *cfxB*, a quinolone resistance gene from a clinical isolate). However, several features of the *soxRS* and *marRAB* regulons are of particular relevance to the general issue of antibiotic resistance, and to the particular situation in the environment: (1) gained "resistance" can add up to a full-resistance phenotype, if coexisting with other low-level resistance determinants, *e.g.*, a *gyr* single-mutation (Heinemann et al., 2000); (2) the activity of different, structurally-unrelated antibiotics is affected by the overexpression of these regulons (*mar* is, after all, an acronym for *multiple antibiotic resistance*); (3) a wide variety of compounds can induce the expression of these regulons, including oxygen and nitrogen reactive species released by activated macrophages (Nunoshiba et al., 1993); several antibiotic (*e.g.*, chloramphenicol, tetracycline (Davin-Regli and Pagès, 2007)) and non-antibiotic drugs (*e.g.*, aspirin (Demple and Amábile-Cuevas, 2003), phenazopyridine (Amábile Cuevas and Arredondo García, 2013)); the lack of iron (Fuentes et al., 2001); and environmental pollutants such as mercury (Fuentes and Amábile Cuevas, 1997) and herbicides (Kurenbach et al., 2015), to mention a few; and (4) mutations in the regulatory genes can result in the constitutive expression of the "resistance" phenotype; such mutations can then be selected by the wide variety of compounds these regulons protect against, including those that act as inducers, and/or other agents against which protection is elicited (*e.g.*, ozone (Jiménez-Arribas et al., 2001) or triclosan (Levy, 2002)). Although protection against oxidative stress can be considered mostly a chromosomal trait, a recent report of a mobile genetic element in *Legionella pneumophila* that confers resistance to hydrogen peroxide and bleach, along with beta-lactam antibiotics (Flynn and Swanson, 2014), opens the possibility of such oxidative stress protective genes to be horizontally acquired.

There is a number of other regulatory proteins that control the expression of efflux pumps in Enterobacteriaceae (*e.g.*, Rob, RamA, PqrA, AarP, AcrR, EmrR), each with its own set of chemical or physical effectors; and a number of efflux pumps other than AcrAB-TolC –37 putative genes in *E. coli* alone (Davin-Regli and Pagès, 2007). The description of the role of each one in adaptive resistance to antibiotics is far from completion. In *P. aeruginosa* the scenario is also complex: the outer membrane porin OprD, and several efflux pumps (MexAB-OprM, MexCD-OprJ, MexEF-OprN, MexXY and MexJK, all of the resistance-nodulation-division, RND family) are involved in multiple antibiotic resistance (Lister et al., 2009). Again, the overexpression of Mex efflux pumps, or the repression of OprD porin, alone, are not enough to confer full-resistance; but when coexisting with other low-level resistance mechanisms, they can add up to a complete protection against clinically-achievable antibiotic concentrations. For instance, diminished expression of OprD, and overexpression of chromosomal beta-lactamase AmpC, can result in carbapenem resistance; and overexpression of MexXY along with aminoglycoside-modifying enzymes can allow *P. aeruginosa* to withstand extremely high concentrations of amikacin. As discussed before, mutations that result either in diminished expression of the porin, and/or increased expression of efflux pumps, lead to a stable multi-resistance phenotype. A number of regulated efflux systems have also been implicated in antibiotic multi-resistance in the opportunistic pathogen *Acinetobacter calcoaceticus-baumannii* (Coyne et al., 2011).

Back to *E. coli*, it is worth to mention that plasmid-mediated efflux pumps known to mediate antibiotic resistance have been described: a conjugative plasmid bearing genes *oqxAB* encode a TolC-dependent efflux system that confers resistance to olaquindox, a bacterial DNA-synthesis inhibitor used as growth promoter in pigs, and also to chloramphenicol (Hansen et al., 2004). By "jumping" to mobile elements, genes that have so far been considered to mediate adaptive or perhaps even intrinsic resistance, could be transferred to other, clinically-relevant bacterial species.

In gram-positives, of course, outer membrane porins are absent; but unspecific efflux systems have been identified, such as the PmrA pump in *S. pneumoniae*, and the Bmr and Blt pumps in *Bacillus subtilis*, whose overexpression reduce the susceptibility to quinolones and other, unrelated compounds (Brenwald et al., 2002). An overview of bacterial antibiotic efflux pumps can be found in Van Bambeke et al. (2003).

1.2.1.1.4 Adaptive resistance? Biofilms

An entirely different set of conditions that enhance the bacterial ability to survive antibiotic exposure result from biofilm formation. Unlike the inducible mechanisms described above, in this case the resistance phenotype is not gained by individual cells, but by the whole biofilm community, by a variety of mechanisms that are yet to be fully understood, and that vary from one bacterial species to the other (Gilbert et al., 2007). Rather than resistance, the word "persistence" seems more adequate: when a biofilm is exposed to antibiotics, many bacterial cells are killed, mainly at the outer layers of the biofilm; but a number of "persisters" are not, making the whole biofilm to survive the exposure and to thrive again afterwards. From the clinical perspective, this means that once the antibiotic treatment is completed, persisting biofilm is still there to grow again, restarting the infectious process; as most infectious episodes are caused by biofilms, and as this phenotype cannot be detected by the usual antibiotic susceptibility assays, it is likely that biofilms are behind therapeutic failure of antibiotic treatments against bacteria deemed susceptible by the clinical lab.

The diminished activity of antibiotics against biofilms is listed here as a form of adaptive resistance, as biofilm formation is often regarded as an inducible, chromosome-mediated ability that allows bacteria to colonize surfaces, hence resistance is only gained when bacteria are growing within such microbial communities. However, biofilm formation itself can be an acquired trait, as conjugative plasmids can mediate biofilm development (Ghigo, 2001). This imply that acquired biofilm-formation traits could also be considered as acquired resistance mechanisms.

In addition to the persistence phenomenon, biofilms and antibiotics interplay at multiple levels: (a) aminoglycoside antibiotics can induce the formation of biofilms (Hoffman et al., 2005); (b) macrolide antibiotics can inhibit the formation of some biofilms (Wozniak and Keyser, 2004), while promoting the formation of others (Wang et al., 2010); (c) biofilm-forming *P. aeruginosa* strains carry more resistance traits than non biofilm-formers (Delissalde and Amábile-Cuevas, 2004); (d) a biofilm can be considered as a significant playground for all sorts of horizontal gene transfer (Amábile-Cuevas, 2013, Amábile-Cuevas and Chicurel, 1996), hence promoting the spread of antibiotic resistance genes, although plasmids seem to be less frequently found in biofilm-forming Vibrionaceae (Xue et al., 2015) and *P. aeruginosa* (Delissalde and Amábile-Cuevas, 2004); and (e) species interactions within biofilms can favor mutations that enable symbiotic, specialized associations (Hansen et al., 2007) that, although not been demonstrated in the context of antibiotic exposure, remain as an intriguing possibility.

1.2.1.2 Co-selection: the plot thickens

Antibiotic usage leds to antibiotic resistance; this we know for sure. But it is not quite as simple: there are non-antibiotic agents that can select for antibiotic resistance; and antibiotics can select traits different from antibiotic resistance. Furthermore, some antibiotics can select for resistance to other, unrelated antibiotics. Most of these interactions are based on co-selection, and have been reviewed before (Amábile-Cuevas, 2013); cross-resistance (*i.e.*, a single resistance mechanism providing protection against several drugs on the same family, or even against chemically unrelated compounds, such as the MLS$_B$ phenotype) that can also account for some of these phenomena, is rather obvious. Important to keep in mind while discussing resistance in the environment are the following:

– Genetic linkage of antibiotic resistance determinants and some other traits, can explain why non-antibiotic agents, or unrelated antibiotics, select for antibiotic resistance genes. For this to happen, resistance genes must reside on the same genetic element: many antibiotic resistance genes have been found along with heavy-metal (*e.g.*, mercury, cadmium) and/or disinfectant (*e.g.*, quaternary ammonium compounds) resistance genes in the same plasmids or other mobile elements. Hence, the presence of such compounds select for the entire genetic element that carries antibiotic resistance determinants, in the absence of antibiotics. It is perhaps relevant to state that the same, or even worst confusion over the definition of "resistance" prevails when referring to disinfectants (Gilbert and McBain, 2003). However, while the role of genes that confer only protection against slightly higher disinfectant concentrations could be negligible in houses or hospitals, it may be particularly relevant in environmental settings where such biocides are diluted. About the same can be said about antibiotics: integrons and transposons (see below) often carry a sulfonamide-resistance gene along with genes conferring resistance to other, unrelated antibiotics. Sulfonamides are among the very few antimicrobial compounds that can be detected at relatively high concentrations in wastewater; therefore, it can select for such multi-resistance genetic elements, in the absence of other, more labile compounds.

– Typical examples of cross-resistance, such as *gyr/par* mutations that protect against almost all quinolones, or *erm* genes that protect against macrolides, lincosamides and streptogramins, are very obvious: the presence of ciprofloxacin would select for ofloxacin or norfloxacin resistance, and the presence of streptogramins would select for clarithromycin resistance. Despite the very simple nature of this assertion, the use of some antibiotics as "growth promoters" circumvented a restriction for using clinical antibiotics by neglecting known cross-resistance: enrofloxacin, while not used in humans, select for resistance to other fluoroquinolones; avoparcin select for resistance to other glycopeptides, such as vancomycin; virginiamycin select for resistance to other streptogramins, such as quinupristin/dalfopristin. Furthermore, other mechanisms of low-level multi-resistance, such as those resulting from unspecific efflux, can be induced by a variety of non-antibiotic agents, and mutants constitutively expressing such mechanisms can be selected by the same kind of compounds.

– By the same token, antibiotics can select for a variety of traits different from antibiotic resistance. Virulence genes have been found linked to resistance ones on the same genetic element, for instance; antibiotics can therefore be selecting for resistant, virulent bacteria. It is even possible that antibiotics can be increasing the prevalence of mobile elements in bacterial populations, either by selecting bacteria that carry resistance plasmids, transposons or integrons; and/or by selecting bacteria that are more permissive of such kinds of extrachromosomal DNA molecules.

1.2.1.3 Inter-molecular gene mobilization: the gene "cut & paste" bacterial kit

Perhaps the most striking feature of the antibiotic resistance crisis is the very high frequency with which multi-resistant (*i.e.*, resistant to three different classes of antibiotics) organisms are isolated in clinical settings. These multi-resistance phenotypes, that are also often transferable in single HGT events, indicate that the accumulation or gathering of resistance determinants in single genetic elements is common. Although the accumulation of resistances in single organisms, due to successive exposure to individual drugs, is merely the consequence of such successive exposure, along with a low rate of spontaneous loss of resistance determinants; the accumulation of resistance genes in single genetic elements, mainly plasmids, is the result of additional phenomena. These involve a number of genetic elements capable of mobilizing between DNA molecules: insertion sequences (IS), transposons, integrons, and gene cassettes, are among the best characterized of them. Although it is not within the purpose of this book to make a detailed review of the nature of each of these elements, there are some important features to highlight.

An old classification of the gene rearrangements mediated by transposons – and extensive to integrons and gene cassettes, put them under the "illegitimate recombination" label; this is to mean that no extensive sequence homology is needed for the insertion of such elements into target DNA molecules, other than short regions (hotspots, attachment sites) that serve as substrate for transposase or integrase enzymes. Through these recombinatorial events, gene cassettes can be inserted, excised and shuffled within integrons; integrons can become linked to transposons; and transposons can "jump" between plasmids, and between plasmids and chromosomes. The result is a sort of Matrioshka doll of nested mobile genetic elements (Amábile-Cuevas and Chicurel, 1992). This picture describe more or less accurately elements such as plasmid R100: a 94.5-kb conjugative plasmid that contains transposons Tn10 and Tn2670 (the later formed by the insertion of Tn21 into Tn9), which in turn contains a class-1 integron, with a single gene cassette inserted; as such, this plasmid, isolated from a *Shigella flexneri* strain in the 1950's, encode resistance to tetracycline, chloramphenicol, sulfonamides, streptomycin, spectinomycin, quaternary ammonium disinfectants, and mercury (Bushman, 2002). Complex arrays of resistance genes can be found in a single integron, as is the case of In53, carrying genes for two beta-lactamases, four aminoglycoside-modifying enzymes, a chloramphenicol-modifying enzyme, along with a *sul1* sulfonamide-resistance gene, and two *qac* quaternary-ammonium compounds' resistance genes (Naas et al., 2001). Some particular details of each of these mobile elements, that are important to understand their role in the spread of antibiotic resistance genes, are:

–   ISs apparently play a minimal role on antibiotic resistance in gram-negatives (*e.g.*, increasing mutagenesis, or inserting promoters upstream silent resistance genes (Amábile-Cuevas, 1993)); but IS257 seems pivotal in the mobilization of resistance genes in gram-positives (Firth, 2003).

–   Transposons seem to mobilize preferentially to plasmids than to chromosomes. This was recognized since the early characterization of these elements, as plasmids were described as "collections" of transposons 40 years ago (Cohen, 1976). About one half of plasmids in a genome database carry at least one IS, with an average density of one copy every 19 kb (contrasting with only 8% of phages), an observation later extended to transposons (Leclercq et al., 2012). Furthermore, some transposons seem to prefer conjugating plasmids as targets for transposition (Wolkow et al., 1996).

–   There are several classes of integrons, but class-1 and -2 are the most commonly linked to antibiotic resistance, and are considered to be mobile, because they are mostly found in plasmids and other mobile elements. Integrase I seems to derive from XerC/D recombinases, having Vibrionales as a sort of bridge towards clinically-relevant enteric bacteria and Pseudomonadales (Díaz-Mejía et al., 2008). While integrons are mostly chromosomal in aquatic bacteria, such as *Vibrio* and *Shewanella*, they are more commonly found in plasmids when in *Pseudomonas* and enterics. Integrons are also frequently found in gram-positives, especially those with similar codon usage, such as *Corynebacterium* (Díaz-Mejía et al., 2008).

–   For gene cassettes to be integrated into integrons, two integrase-specific recombination sites are necessary: an *attI* site at the integron, and an *attC* site at the cassette. Most curiously, gene cassettes are composed of a single gene and the *attC* site, suggesting that reverse transcription of mRNA molecules are at the origin of such elements. Group IIC introns have been proposed as responsible for the formation of gene cassettes, providing the *attC* site to unrelated genes, and the retrotranscription machinery to form the cassette (Léon and Roy, 2009). A class 1 integron bearing a group II intron was found in an *E. coli* from a wild Norway reindeer (Sunde, 2005), indicating that this conjunction likely existed before and without antibiotic intervention.

- Transposition and integration, in several instances, are increased when the host cell has activated its SOS response. From the early reports of the transposase of IS*50* and derived composite transposons (*e.g.*, Tn*5*) being repressed by LexA (Kuan and Tessman, 1991); to the more recent discovery of induction of integrons' integrases by an activated SOS response, which fosters the acquisition and rearrangement of gene cassettes (Guerin et al., 2009). Antibiotics themselves play a role in inducing the mobilization of transposons and integron gene cassettes (Courvalin, 2008).

Through transposition and integration, resistance genes can be mobilized between DNA molecules residing within a single bacterial cell. Such rearrangements can allow a better expression profile of gene cassettes, "adequating" it to the environmental conditions; can be fostered by the exposure to environmental stress; and allow the assembly of complex, nested, multi-resistance mobile elements. But the substrate genes for such formidable arrangements come from very different cell lineages, as has been demonstrated by sequence homology and codon usage. Therefore, these mechanisms alone may have been meaningless without the ability to exchange genetic information between cells: the horizontal gene transfer.

1.2.1.4 *Horizontal gene transfer: the main means for resistance spread*

Horizontal gene transfer (HGT) seems to be a peculiar feature of prokaryotes: 30–50% of Bacteria have at least one protein domain acquired through HGT, while less than 10% of Eukarya do (Choi and Kim, 2007). The extent of this transfer establishes a "network of genomes" (Dagan and Martin, 2009) among bacteria, of a magnitude and consequences difficult to grasp. If an old calculation of foreign DNA in the genome of *E. coli*, of about 12.8% (Ochman et al., 2000), still holds, and if it is somehow representative of similar bacteria, it is at least easy to realize that HGT is quite common. There are three main mechanisms of HGT (transformation, transduction, and conjugation), each with a number of theme variations; but, before briefly reviewing these mechanisms, lets enlist some traits that are known to have been mobilized in these ways.

Of obvious relevance for the purpose of this book, hundreds of genes directly mediating antibiotic resistance are known to have been mobilized horizontally. But resistance genes have arguably been caught in this gene flux only recently, as the human use of antibiotics mounted a sudden and dramatic selective pressure. Aside from resistance genes, it is generally accepted that not all kinds of genes are equally transferable, establishing an initial bias as to the kind of genetic information that can be exchanged. For instance, genes that are part of complex systems, such as transcription or translation (named "informational genes") are less frequently transferred than those that act relatively on their own mediating housekeeping traits ("operational genes") (Jain et al., 1999). Within these housekeeping genes, those coding for secreted

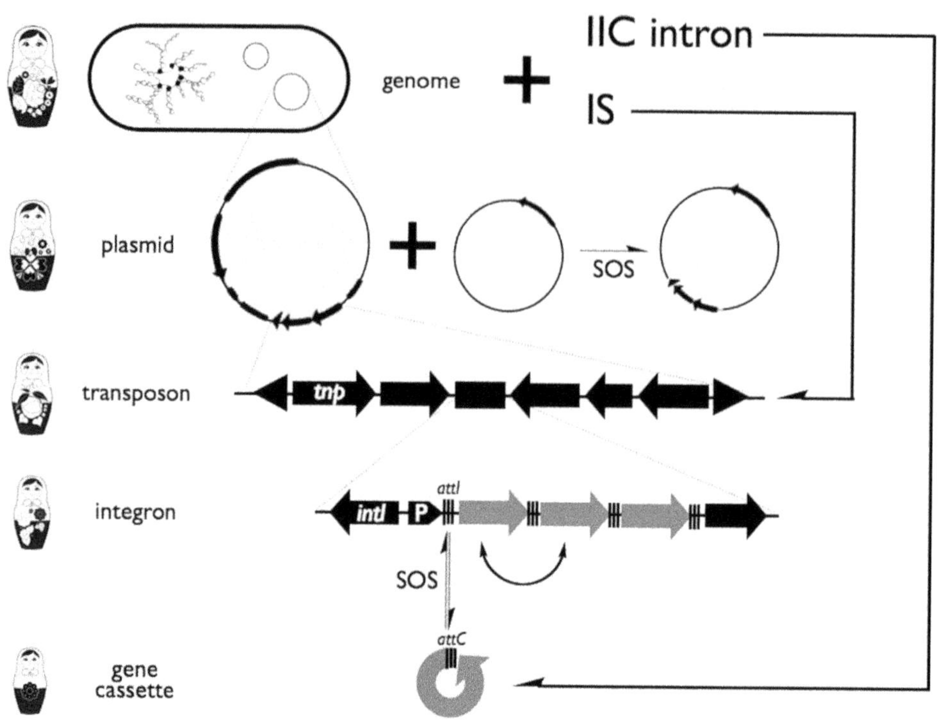

Figure 1.3 **Intra-cellular gene mobilization; a graphic summary.** A series of genetic elements abound within bacterial cells, many in a nested fashion resembling a Matrioshka doll. The whole genome of a bacterial cell is formed by one (or two) chromosomes, and one or several plasmids. Plasmids, usually ranging from 1 to 10% of the size of the chromosome, can be seen as "accessory" gene elements, or as sub-cellular forms of life. They can act as collections of transposons, that can go from the simple insertion sequence (IS, containing only genes for transposase and resolvase, and substrate sequences for transposase), to very complex sets of genes, including antibiotic resistance ones. Some transposons include other kind of element, the integron, that is essentially formed by an integrase (*intI*) gene, a promoter sequence, and an attachment site (*attI*); the integrase allows the recombination between *attI* site of the integron, and the *attC* site of a gene cassette, a non-replicative DNA circle containing a single gene and an *attC* sequence. SOS responses can induce transposition and integration, allowing the mobilization of transposons between coresiding DNA molecules, and/or the acquisition, excision, or rearrangement of gene cassettes. Composite transposons can result from the insertion of ISs flanking a gene or set of genes; and gene cassettes from group CII introns and their associated retrotranscription activities.

proteins and for outer membrane proteins seem to be more prone to mobilization than those that encode periplasmic, cell membrane, or cytoplasmic proteins (Nogueira et al., 2009). Furthermore, most analysis of mobilized genes are centered on protein-coding sequences; but a recent paper suggests that there is a sort of regulatory "switching" that can explain the expression divergence between strains; and that such switching occur through HTG of regulatory regions (Oren et al., 2014). With regulatory genes also included in the gene pool formed by HGT, and adaptive resistance included in the genome of many bacterial types, the acquisition of resistance phenotypes has yet another potential avenue to arise.

It is important to realize that, although grouped under the same "HGT" category, HGT mechanisms differ widely, and that these differences do have an impact upon the kind of genes that are preferentially mobilizable. Conjugative transfer, that depends on dedicated mobile genetic elements, as will be reviewed below, is most likely to mobilize traits that are involved in the social behavior of bacteria, as such genes can provide a competitive advantage for the host cell (Rankin et al., 2010). In contrast, mobilization via phages can be regarded as "accidental"; and transformation, that mostly depends on ulterior recombinatorial events for the successful gene acquisition, can be seen as a sort of DNA repair mechanism (and a source of subtle gene variation), as will be further discussed. Antibiotic resistance genes can be set apart: they provide a decisive selective advantage, and can function as a sort of "post-segregational killing" mechanism (present in some plasmids and enhancing their stable inheritance (Gerdes et al., 1986)), acting the "antitoxin" role, with the "toxin" being the external presence of antibiotics (Heinemann and Silby, 2003).

One of the many controversies regarding the extent and impact of HGT is directly related to the environment. Some authors consider that for HGT to happen, a main influence shaping the "network of gene exchange" is a shared ecological niche (Smillie et al., 2011). It makes sense to imagine that, for genes to flow from one organism to another, such organisms must be in proximity, which means a shared environment. But others find that the main factor influencing said network is phylogeny (Skippington and Ragan, 2012), with closely related bacteria participating in HGT at higher rates, regardless of environment or lifestyle. It also makes sense that, for horizontally-acquired genes to be successfully established and expressed in new hosts, donor and recipient cells must be related (e.g., by similar codon usage (Medrano-Soto et al., 2004)). It is likely that there is not a single answer for all bacterial species and HGT mechanisms; it is also likely that antibiotic resistance genes can overcome any and all limitations for HGT, as they provide protection against a lethal force.

From the clinical point of view, along with antibiotic resistance, many virulence genes can be acquired via HGT. Virulence traits fall within the typical mobilizable genes, as described above. It is particularly worrisome to think of virulence and resistance genes residing in the same mobile element, as has been documented a number of times (e.g. (Guerra et al., 2002)). But from the biological point of view, it is perhaps much more worrisome to think of the "evolvability of evolvability" (Koonin and Wolf, 2012), and to link it to HGT and antibiotic resistance. Evolvability is often considered as a consequence of mutation, and error-prone DNA repair mechanisms may have a significant role in it. Hypermutability has been related to antibiotic presence and resistance in many different ways (Martínez and Baquero, 2000); and error-prone repair mechanisms have been reported to be horizontally mobilized (Brown et al., 2001), even within pre-antibiotic era plasmids (Sedgwick et al., 1989). Therefore, it is easy to draw a theoretical line connecting evolvability (error-prone repair, hypermutability) to HTG (conjugative plasmids) to antibiotics; and to hypothesize on the role of antibiotics as a selective pressure, or perhaps even inducing factor, not only for antibiotic resistance genes, but also for evolvability itself. However, the links could be much deeper than that: (a) as the acquisition of resistance genes via HGT is a clear advantage in the presence of antibiotics, antibiotics may be selecting organisms that are more prone to HGT (the "survival of the best connected" (Baquero, 2011)), hence fostering evolution; (b) as HGT is enhanced by close proximity between donors and receivers, mobile elements may be recruiting mechanisms that promote such proximity (that include virulence traits (Amábile-Cuevas and Chicurel, 1996)), and antibiotics may be co-selecting for these mechanisms if linked to resistance genes; and (c) as protection conferred by antibiotic resistance mechanisms can extend beyond the resistant cell lineage (*e.g.*, extra-cellular antibiotic-inactivating enzymes), and antibiotics themselves can increase HGT in a number of ways, HGT between resistant and susceptible bacteria may be promoted by antibiotics' presence in different settings.

The peculiar, intertwined role of the SOS system in HGT and antibiotic resistance deserves a separate paragraph, especially when analyzing these issues and the environment. Essentially considered to be a response to DNA damage, a large number of physical and chemical conditions that bacteria in open environments can encounter can trigger this response; furthermore, additional situations that do not seem to involve DNA damage can also activate the SOS response (*e.g.*, high pressure (Aertsen et al., 2004), defective cell wall synthesis caused by beta-lactams (Miller et al., 2004)). The SOS response in itself has been singled out as responsible for causing mutations leading to resistance to quinolones and rifamycins; both antibiotic classes are also known SOS inducers (Cirz et al., 2005). Furthermore, as was described in the section above, and will be analyzed in this one, the SOS response is also induced and/or activates inter-molecular (transposition and integration) and inter-cellular (transduction and conjugation) gene mobilization. It is therefore of particular importance to keep in mind that all environmental conditions that activate an SOS response might foster the emergence and spread of antibiotic resistance.

As with antibiotic resistance mechanisms, there are also many, many reviews on HGT mechanisms in bacteria; an old paper (Amábile-Cuevas and Chicurel, 1992) along with an "update" (Amábile-Cuevas, 2013) can provide a reasonable overall background on the issue. Although the goal of this book is to focus on antibiotics and antibiotic resistance in the environment, it is impossible to do so without understanding the role of HGT. In the following paragraphs, an overview of HGT mechanisms, with special emphasis on its occurrence in open environments, will be provided.

1.2.1.4.1 Transformation

Transformation, first described in pneumococci, is the result of the uptake of free DNA (with an exception discussed below) by bacterial cells able to do so, an stage called "competence"; and the further incorporation of said DNA into the bacterial genome. Allele exchange leading to variation in surface markers in order to avoid immune response, is a consequence of transformation in the human pathogen *Neisseria gonorrhoeae*. Relevant conditions and mechanistic details of transformation include:

–   Transforming DNA, in the original experiments by Griffith and Avery's team, was released from bacteria by heat-killing, something perhaps not likely to happen in nature (heat-killing occurs, of course, but such an event would also kill potential recipients, in natural scenarios). Bacterial autolysis can naturally release transforming DNA; additionally, some bacteria secrete DNA through type-IV secretion systems (T4SS). While the DNA secretion phenomenon was described many years ago, its "purpose" remains a mystery: it may be a deliberate means for DNA exchange, or perhaps only serving other functions, such as biofilm formation (Tetz et al., 2009). Hydrogen peroxide can also induce the release of DNA: in *Streptococcus gordonii*, enzymatic production of H_2O_2 causes both, DNA release and competence induction (Itzek et al., 2011).

- A peculiar mode of DNA "secretion" is the packing of DNA into outer membrane vesicles by a number of bacterial genera (*e.g.*, *Acinetobacter, Pseudomonas, Neisseria, Escherichia*). Such vesicles function as toxin delivery vehicles, along with other roles in pathogenesis; but they also carry DNA that can be uptaken by other bacterial cells. Virulence genes are known to be transferred through these vesicles, but there are some reports of resistance genes also being mobilized in this way: chromosomal beta-lactamase genes have been found in vesicles of *P. aeruginosa* (Ciofu et al., 2000); and carbapenemase-encoding plasmids of *A. baumannii* were readily transferred from one strain to another through vesicles(Rumbo et al., 2011). This form of "transformation" circumvents some of the problems often associated to free-DNA transformation: vesicle-contained DNA is protected from enzyme degradation; and it seems to allow for large, double-stranded DNA to be mobilized, perhaps even without the need for specific sequences or further homologous recombination (see details below).

- Two artificial sources of transforming DNA, especially important for the purposes of this book, are the breakdown products of antibiotic-producing bacteria present in antibiotic preparations; and DNA from transgenic plants that include antibiotic resistance genes as remnants of the early processes of DNA manipulation. Bacterial DNA was detected, by PCR amplification of 16S rRNA genes, in a number of antibiotic preparations of aminoglycosides, tetracyclines, glycopeptides, cephamycins, and macrolides; amplifiable resistance genes were found in streptomycin and oxytetracycline formulations (Webb and Davies, 1993). As with many of the phenomena related to antibiotic resistance, the likelihood of such events to occur can be very low (even below our detection capability), yet they do happen, due to the formidable number of bacterial cells and their rapid reproduction. Therefore, although a brief communication tried to discard antibiotic preparations as sources of resistance genes in clinical isolates (based on the analysis of the few sequences available in databases by 2004 (Lau et al., 2004)); and not all antibiotic formulations seem to carry resistance genes (Rezzonico et al., 2009); substantial contamination with bacterial DNA was found in the crude antimicrobial formulations used as animal-feed additives, including the cluster of glycopeptide-resistance genes, present in avoparcin preparations (Lu et al., 2004). As for transgenic plants, DNA from decaying leaves survives the physical and biological degradation inside the plant tissues, being in contact with transformable saprotrophs, and can even reach the soil. This was demonstrated for the *aadA* gene, conferring streptomycin resistance, used in the construction of transgenic tobacco plants (Ceccherini et al., 2003, Pontiroli et al., 2009). While these experiments used recipient bacteria already carrying defective resistance genes, which were then replaced by homologous recombination, a set of circumstances difficult to find naturally; they offer a proof-of-concept for the dispersal of genetic information from genetically-modified organisms via transformation in the environment.

- Only a few bacterial genera are known to achieve transformation competence naturally; of direct clinical relevance are *Acinetobacter, Haemophilus, Legionella*,

Mycobacterium, Moraxella, Neisseria, Pseudomonas, Streptococcus and *Vibrio* (in alphabetical order). Also interesting, for the potential bridging between the soil environment, where they are abundant, and the clinical one, is the genus *Bacillus*; and *Streptomyces*, the antibiotic-producing genus, that achieve competence and also secretes DNA. Of course, there could be many more, unknown bacterial genera capable or achieving competence; and perhaps even naturally-occurring conditions that may induce competence in bacteria that cannot achieve it on their own, in ways similar than the ones used in the lab for *E. coli*.

– DNA uptake is, in some of the well-characterized models of transformation, very sequence-specific, leaving little room for the acquisition of genetic information from distantly-related bacteria. In *N. gonorrhoeae*, for instance, DNA binding leading to uptake only happens when a 10-base sequence, frequently found in the chromosome of related bacteria, is present (Hamilton and Dillard, 2006); and uptaken DNA is single-stranded (and one strand is preferentially uptaken, as was reported in the only paper known to this author using the "Watson" and "Crick" strands denomination (Duffin and Seifert, 2012)). Something similar happens in *S. pneumoniae* and *Haemophilus influenzae*. The *B. subtilis* laboratory strains do not seem to be so sequence-selective, therefore completely foreign DNA can be used for transformation; however, wild strains are much more difficult to transform by natural competence (Nijland et al., 2010), indicating that some of the things we have learned from lab strains might not apply to environmental isolates. *Acinetobacter* and *Pseudomonas* seem to have little or no sequence specificity for initial binding and uptake, but homology is crucial for further recombinatorial incorporation of acquired genetic information (de Vries et al., 2001).

– Competence can be achieved in many different ways: some bacteria are constitutively competent, such as *N. gonorrhoeae*; in some others, competence is inducible by external factors, often quorum signaling molecules, as in *Bacillus* and *Streptococcus*; and yet in others, competence is achieved as a physiological stage dependent on internal factors, as in *Haemophilus* and *Pseudomonas*. A report of natural transformation in *Streptococcus mutans* include a quorum-sensing pheromone inducing competence, and a 10- to 600-fold increase in transformation efficiency when cells were growing in a biofilm (Li et al., 2001). Genotoxic stress induce competence in *Legionella pneumophila* (Charpentier et al., 2011).

– It is generally assumed that transforming DNA must be released at a short distance from competent bacteria, and shortly before being uptaken; otherwise, free DNA will be quickly degraded. However, a significant amount of extracellular DNA can be recovered from soils (Pruden and Arabi, 2012), indicating that release outpaces degradation. Furthermore, clay and other environmental components can protect DNA from degradation, extending the "transforming range" of free DNA in the environment. DNA-containing vesicles, discussed above, may also have extended half-lives in the environment.

1.2.1.4.2 Transduction

Transduction is a rather "accidental" HGT mechanism, in which genetic information from a bacterial cell, infected by a bacteriophage, is carried by the offspring viral particles into newly infected bacterial cells. Transduction has been traditionally

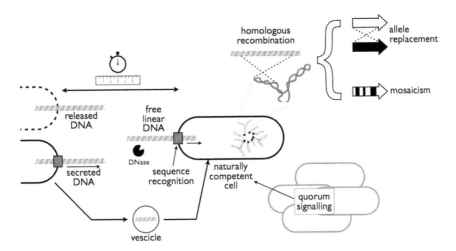

Figure 1.4 **Transformation; a graphic summary.** Transformation requires the uptake of free DNA by competent bacteria. Only a few bacterial genera are known to achieve competence naturally, some times as a response to quorum signaling. The DNA uptake machinery recognize sequences from related species as substrate, hence limiting the origin of DNA to be uptaken to such related species. Free DNA can come from dead bacteria (as in the original experiment by Griffith in the 1920's); or by the secretion of DNA by some bacteria. It is assumed that such free DNA must be released very close, in time and distance, to competent bacteria, or else the DNA is degraded. An alternative method involves membrane vesicles containing DNA from donor bacteria; in this way, DNA is protected from degradation. Once inside the cell, acquired DNA undergo homologous recombination with native sequences, leading to the replacement of whole genes, or to "patch" replacement, or mosaicism. Rarely, whole plasmids can enter a competent cell and become established as independent replicons, although plasmids in vesicles are known to be efficiently uptaken.

classified as (a) specialized, when the mobilized bacterial gene is one adjacent to the chromosomal insertion site of the phage during lysogeny, and an aberrant excision of the phage DNA at the beginning of the lytic cycle results in a hybrid phage-bacterial DNA molecule; and (b) generalized, when a piece of DNA, usually of the same size as the phage genome, is mistakenly enclosed in a phage capside during the lytic cycle of the phage; such piece can be a chromosomal fragment or even a plasmid. However, many phages carry, within their standard genome, genes that can hardly be considered of phage origin. Mobilization of such genes occur each time the temperate phage carrying them switch to lytic cycle, and cannot be considered either specialized nor generalized transduction. There is plenty of information of bacterial virulence genes included in phage genomes: perhaps the most studied example is the diphtheria exotoxin, produced only by lysogenic strains of β and ω phages that infect *Corynebacterium diphtheriae*; only lysogenic strains are virulent. There is also, however, growing evidence of antibiotic resistance genes being mobilized by phages (Colomer-Lluch et al., 2011). Further details to be considered around transduction are:

– Many, but not all phages, display species specificity – some are only capable of infecting particular strains, or phagotype, of a bacterial species. This limits the range of spread of transduction as means for HGT, depending on the infectivity range of the carrier phage.

- Different conditions can trigger the switch from lysogenic to lytic cycle, actually ensuing HGT; among such conditions, for some phages, is the SOS response, that can be activated by environmental agents, including some antibiotics. This is a clear instance where antibiotics can induce HGT.
- Phages are utterly abundant: an old study found 2.5×10^8 virus/mL in natural waters (Bergh et al., 1989). A more recent paper found around 1,000 different phage and prophage genomes in human fecal samples, out of nearly 4,200 viral genomes present there (Reyes et al., 2010). This is to say that, despite a low probability of phage-mediated HGT, the numbers and variety of phages in the environment make the phenomenon inevitable. A recent report of transduction by "particles" of apparently infectious nature and found in different environments (Chiura et al., 2011), adds to the potential frequency of this kind of HGT.
- An old, Soviet-time concept of phage therapy (Sulakvelidze et al., 2001), is gaining traction as an alternative means to combat antibiotic-resistant bacteria. The use of phages in farm animals is now happening (Kittler et al., 2013). Although the concept is interesting and has shown clinical efficacy, the environmental consequences of releasing huge amounts of therapeutic phages, that can act as HGT vehicles, has not been adequately explored (Meaden and Koskella, 2013).

1.2.1.4.3 Conjugation

The main mechanism of HGT in bacteria is conjugation. The complex conjugative machinery is always encoded by mobile genetic elements, which range from symbionts to parasites of the bacterial host; therefore, conjugative elements are in many ways independent of the fate of the host. Once a recipient cell gains the conjugative element, it often turns into a donor cell, having thus a cascade effect. Conjugation-mediated HGT has the widest known range, including distantly related bacteria, archeobacteria, and many eukaryotic cells, from yeasts, to plants, to human gut cells. Mechanistically, conjugation depends on cell-to-cell contact, that is achieved in a variety of ways, from the retractable F pilus of *E. coli*, to the pheromone-inducible aggregation substance of enterococci (intrachain conjugation, *i.e.*, mobilization of conjugative elements along chains of bacteria, such as those formed by *Bacillus subtilis*, was recently reported (Babic et al., 2011)). Once contact has been established, a pore containing a T4SS enables the transfer of single-stranded DNA molecule (bound to a relaxase molecule, attached to an *oriT* site) from donor to recipient cells, along with several proteins that protect the ssDNA molecule from attack by donor's immunity. The transferred DNA molecule then recircularizes, still as single-stranded, and the replication machinery of the cell synthesizes the complementary strand – also in the donor cell. Although conjugative plasmids were thought to be the most common of elements transferable by conjugation, a rather new denomination of ICEs (integrative conjugative elements, formerly known as conjugative transposons, that are unable of independent replication and must therefore insert into the host's chromosome using different biochemical strategies) seem to be more abundant. Conjugative elements encode the whole machinery for DNA transfer, but other, smaller elements can be mobilized along, provided they have a compatible *oriT*; they are the mobilizable plasmids and IMEs (integrative

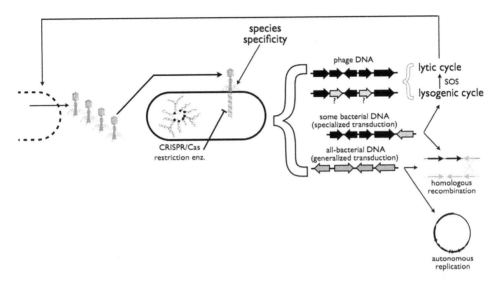

Figure 1.5 **Transduction; a graphic summary.** Transduction depends on a phage vector to mobilize DNA; phages are released by infected bacteria and infect other cells. Many phages can only infect a specific bacterial species (or even a specific variety), limiting the range of mobilization through transduction. "Injected" DNA must override immunity systems, such as CRISPR/Cas and restriction enzymes, in order to continue their cycle. This DNA can be solely of phage origin (or permanently included genes of likely bacterial origin); depending on the phage and cellular conditions, the phage can then insert into the bacterial chromosome (lysogenic cycle) and be vertically inherited; or can directly go into lytic cycle. SOS responses play an important role in turning from lysogenic into lytic cycles. When an inserted phage genome is wrongfully excised, it can "pick up" a bacterial gene, resulting in specialized transduction. If the phage loses its replication abilities due to the aberrant excision, the bacterial gene can undergo homologous recombination with a native gene. If an entirely bacterial DNA fragment gets enclosed into a phage capside (generalized transduction), such DNA can also undergo homologous recombination or, if it is a whole plasmid, can establish itself as an independent replicon. There are "hybrids", *i.e.*, DNA molecules that live independently as plasmids, but mobilize themselves as phages.

mobilizable elements), which in turn are more abundant than conjugative plasmids and ICEs (Guglielmini et al., 2011). Some further features of conjugation that will become relevant for the purposes of this book are:

–   Conjugation can occur in a variety of conditions, including within intracellular compartments (Ferguson et al., 2002); also, some conjugative plasmids can turn non-viable cells into conjugative donors, as they encode all necessary functions for DNA mobilization (Heinemann and Ankenbauer, 1993b). This include cells under the influence of bacteriostatic antibiotics (Cooper and Heinemann, 2000).
–   The notion of retrotransfer, *i.e.*, the transfer of DNA from "recipient" to "donor" cells during conjugation, was entertained during the 1990's; however this was shown to happen as a two-step process: in a first conjugative event, the donor transfers a conjugative plasmid to the recipient, turning it into a donor; in a second

conjugative event, a recipient-turned-donor transfers DNA to a donor-turned-recipient (Heinemann and Ankenbauer, 1993a). This is particularly relevant as a partial potential answer to an old question: what is there for donors to gain by conjugation?

- Conjugative mobilization can be induced by an activated SOS response: an ICE from *Vibrio cholerae*, conferring resistance to several antibiotics, increases the expression of mobilizing genes and the frequency of conjugative transfer by SOS activation which, by the way, was induced by ciprofloxacin exposure (Beaber et al., 2004); and by receiving a ssDNA molecule during conjugation, an SOS response can be activated by conjugation which, in turn, can induce the mobilization of integron gene cassettes within the recipient cell (Baharoglu et al., 2010).

- Although conjugation functions are encoded by the conjugative element, host factors also play a role in the spread of the mobile element. Efficient donors within bacterial populations can mobilize conjugative plasmids very quickly and to many recipients, exerting a sort of amplification effect (Dionisio et al., 2002). There is little information on the nature and presence of such efficient donors in the environment, and the selective pressures that may favor them.

1.2.2 Detection of resistance in the environment

As discussed above, "resistance" can mean many different things, hence its detection is necessarily limited and confuse. There are two main means for assessing the presence of antibiotic resistance in the environment: culture-based and molecular-based methods. Both can provide relevant, often complementary information; both have important limitations; and both can lead to many levels of misinformation (Figure 1.7). Many recent studies try to mix these techniques, as when molecularly detecting relevant genes in environmental isolates, or when introducing isolated genes into live bacteria to detect resistance phenotypes. None of these approaches provide complete answers, as will be discussed below.

Culture-based methods rely on the growth and further characterization of bacteria from environmental samples, such as soil, water, or wildlife feces. The main limitations of this approach are the very small proportion of bacterial species that can actually be grown in available culture media, which some authors place in around 1%; and the difficulties in detecting resistant bacteria that might be present, but in very small quantities, requiring the processing of large samples. These problems cannot be avoided, and all results from culture-based surveys must consider such limitations. Further potential problems can be ameliorated, by better designing of experimental conditions:

- Some surveys use selective plates (*i.e.*, agar plates containing antibiotics, so that only resistant bacteria can grow) to look for resistant bacteria, inoculating the sample directly or, better yet, after an enrichment process. This strategy further limits the reach of the survey, as only a few kinds of culture media can be used along with antibiotics (*e.g.*, Mueller-Hinton agar); and antibiotic concentrations are arbitrarily established, as even clinical breakpoints could be meaningless for environmental bacteria. In order to use other media that could increase the recovery of different bacterial species, it would be necessary to test for possible interference with antibiotic activity. It would also be advisable to use low concentrations of

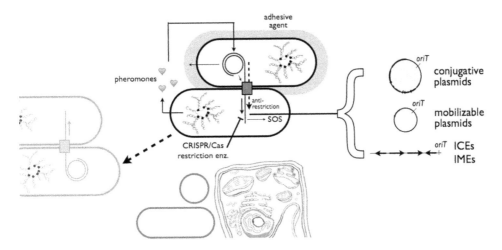

Figure 1.6 **Conjugation; a graphic summary.** Conjugation depends on the close proximity of a donor cell (top) and a recipient cell (bottom). Such proximity results from adhesive agents, typically an F pilus, in gram-negatives, or an adhesin covering the cell, in gram-positives. Adhesive agents are encoded by the conjugative element. In the particular case of enterococci, recipient cells produce pheromones that induce the expression of the conjugative machinery in the donor; an especial case of an active role of the recipient cell. Single-stranded DNA is mobilized from the donor to the recipient, often accompanied by anti-restriction mechanisms; recently-acquired DNA must face CRISPR/Cas (although, apparently, not as efficient against plasmids as it is against phages) and restriction enzymes. Also, as DNA enters as a single-stranded molecule, it activates SOS responses. Elements mobilizable by conjugation include conjugative plasmids, that encode the whole conjugative machinery; mobilizable plasmids, that depends on the conjugative machinery encoded by co-resident conjugative elements, and have *oriT* regions compatible with such co-resident element; or a variety of integrated chromosomal elements (ICEs), formerly known as conjugative transposons, that encode the conjugative machinery but are not independent replicons. Conjugation is a wide-range phenomenon: it can occur between very distantly related bacteria, and between bacteria and a variety of eukaryotes, including human gut cells. A distinctive feature of conjugation is that the former recipient usually becomes a donor; this allows, among other things, a sort of "retro-transfer", *i.e.*, the mobilization from recipient to donor, once thought to happen during the same conjugative event.

antibiotics, and then discard intrinsically-resistant bacteria; rather than use high concentrations and miss important low- or mid-level resistance phenotypes.

- Intrinsic resistance pose a significant handicap for environmental surveys of resistance; it is very common to read reports of "high prevalence of resistance", that are contaminated by a number of intrinsically resistant bacterial species: *e.g.*, enteric bacteria resistant to macrolides, *Pseudomonas* resistant to cephalosporins (other than ceftazidime), etc. While intrinsic resistance is often known for clinically-relevant species, it is an obscure area for most environmental bacteria. It would be advisable to learn more about the nature of a resistance phenotype in previously uncharacterized species, before sounding the "resistance" alarm. If the prevalence of a resistance phenotype in a given species is very close to 100%, it is most likely an intrinsic trait.

– As with many clinical surveys, the adequate choosing of the antibiotics to be tested can avoid redundant or useless information, while providing additional insights about the underlying mechanisms of resistance. If using automated susceptibility assays, there is little room for choosing the antibiotics to be tested; but if using disks and/or in-house dilution assays, it is possible to get much more information than the simple presence of a resistance phenotype. Some examples of typically redundant or useless assays in environmental surveys include: (a) testing several fluoroquinolones against gram-negatives, as cross-resistance is a rule; (b) testing streptomycin against enteric bacteria, as it is never used clinically for this purpose, and resistance is often the result of non-transmissible ribosomal mutations (streptomycin resistance can be enzymatically-mediated, and such an enzyme is typically encoded by a gene cassette; but then it would be much more informative to run tests for the presence of integrons); (c) testing antibiotics against intrinsically-resistant bacteria. On the other hand, running susceptibility towards amoxicillin-clavulanate, when using disks, can reveal the plasmidic or chromosomal nature of the beta-lactam resistance (*i.e.*, ampicillin and amoxicillin-clavulanate resistance is often chromosomal, while ampicillin-resistance, amoxicillin-clavulanate susceptibility is more likely plasmid-mediated); and placing the amoxicillin-clavulanate disk close to a third-generation cephalosporin disk, could reveal the presence of ESBL. Disk-based assays can also distinguish the biochemical mechanisms underlying carbapenem and macrolide resistance.

Molecular methods have evolved rapidly, from the PCR-detection of known resistance or mobility genes, to functional metagenomic studies. By directly detecting DNA sequences on environmental samples, these methods circumvent the need for culturing, being therefore capable of detecting resistance genes in non-culturable organisms. Furthermore, the sensibility of many of these methods is very high, enabling the detection of minute quantities of genetic information within rather large samples; by adding quantitative features to such techniques, it is even possible to know how many copies of a given gene are present. Early methods relayed on PCR, using primers of known genes to detect their presence in environmental samples. The limitations of such methods are clear: (a) only known genes can be detected; (b) minute variations in the sequence used for priming can lead to a functional gene to be missed (false negatives); (c) non-functional resistance genes having intact priming sequences could yield false-positives, if amplicons are not subsequently sequenced; (d) bacteria that are resistant to lytic pre-amplification procedures could hide resistance genes that go undetected; (e) there is no easy way to link the detected gene with its carrier organism,; and (f) resistance genes that are present but not expressed could lead to misconceptions regarding environmental selective or maintenance pressures (although pointing at potential reservoirs). An interesting modification, by looking for genes linked to the resistance one, was able to detect mobility determinants, hence allowing, to some extent, to know how likely the resistance gene is to be subject of HGT. Despite all limitations mentioned above, PCR-based detection is still widely used and provide useful bits of information, such as the presence of integrase genes, and specific resistance determinants in species that are of particular interest, as well as when trying to track down the spread of resistance in the environment. The emergence of micro-arrays just made detection of multiple, known resistance genes much more expedite, but always within the limitations mentioned above.

Metagenomics was a game-changer for the search of resistance in the environment, mostly by assessing the real extent of the unculturable microbiota and revealing the main limitation of culture-based methods for detecting resistance in the environment. Also, by allowing the comparison of found sequences with databases (see below) of resistance determinants, it was possible to detect far more resistance genes than it would have been practical using PCR-based methods. However, the reach of these methods rapidly exceeded their grasp, when researchers started to report as "resistance genes" sequences found in ill-curated databases, or without full identity to reported determinants, or even "putative resistance genes", based on the predicted activity of a predicted protein. Also, these methods were unable to identify the nature of the carrier organism, or if there was actual resistance, phenotypically speaking, as a result of such gene carriage. A recent example of this speculation-prone field is the report of resistance in the microbiota of members of the Yanomami Amerindian village, in Venezuela, living in isolation and without exposure to antibiotics. While none of 131 E. coli isolates were resistant to any of the 23 antibiotics tested (but carried "resistance genes targeted against eight antibiotics"), a metagenomic approach detected a PBP naturally insensitive to cephalosporins, among 28 other "functional antibiotic resistance genes"; the conclusion: "despite different antibiotic exposures, the microbiota from antibiotic-naïve and industrialized people share a common resistome"! (Clemente et al., 2015). Functional metagenomics (i.e., cloning extracted DNA into plasmids, transforming host bacteria, usually E. coli, with such plasmids, and looking for conferred resistance phenotypes) was able to partially fix the bold claims of "resistance" based on sequence homology and/or structure-function analyses; but introduced its own set of speculations: taking a gene, completely out of cellular context, into a lab strain, and detect "acquired resistance", is also a very bold, and potentially wrong statement. Furthermore, "false negatives" can occur much more frequently, as expression problems (e.g., lack or promoters, different codon usage and/or post-traductional modifications) can hinder the detection of an actual resistance gene in a heterologous host. The further coupling of proteomics to functional metagenomic approaches has been recently proposed (Fouhy et al., 2015) as means for enhancing the detection capacity of these techniques.

And then, there is the problem of resistance gene databases. There seems to be a divorce between bioinformatics and microbiology during the assembly and curation of some available databases. The result is that, along with genes that specifically mediate full-resistance to antibiotics, that have been or potentially can be mobilized by HGT, such as those encoding inactivating enzymes (e.g., beta-lactamases) of specific-efflux pumps (e.g., tet or mef genes); there are genes that mediate adaptive, unspecific low-level resistance, such as those for AcrAB-TolC efflux systems; determinants for intrinsic resistance, such as some penicillin-binding proteins; or genes that do not act individually to confer resistance, but need the concerted action of several others (e.g., glycopeptide resistance operons). By indiscriminately placing all these genes under the "resistance" label, metagenomic experiments can dramatically overestimate the presence of resistance and the risks attached. As a result, people was initially surprised to learn about the omnipresence of "resistance" genes, and then dismissed relevant findings of resistance because "resistance is well-known to be everywhere".

One of the best efforts to have all these nuances considered in the curation of the database is the CARD (Comprehensive Antibiotic Resistance Database; (McArthur et al., 2013)). Still, the database includes mutant porin proteins, regulatory genes for beta-lactamases and PBPs, unspecific efflux pumps, and topoisomerase mutants, that are all very unlikely to be horizontally transferred, and many are responsible for intrinsic or adaptive resistance. Therefore, by simply matching DNA sequences found in metagenomic experiments, to genes in the CARD, without going into the annotations included in the database, would lead again to an overestimation of the presence of resistance genes in the environment. This led to a further framework for risk assessment of resistance genes found by metagenomic analysis: the RESCon, or "readiness condition", which the will assign risk categories depending on the perceived likelihood of such genes to become a health problem (Martínez et al., 2015). This will be discussed in the next chapters.

We have certainly learn a lot from the first assessments of resistance in the environment, especially regarding the limitations of each strategy. It seems clear that neither culture-based not molecular-based approaches can provide a full picture of the problem; and it is also evident that multi-disciplinary teams must be assembled to combine all the areas of expertise that are required for an adequate analysis of resistance. These go from the clinical knowledge of relevant antibiotics and concentrations, to the basic microbiology required for deciding culture media and for distinguishing intrinsic resistance, to the molecular biology of resistance determinants and HGT, to the bioinformatic arena that compiles and analyses the metagenomic data.

1.3 ENVIRONMENT

The word "environment" can be used in a wide variety of ways; therefore, it is important to set limits to what kinds of "environment" would be analyzed in this book. When administering antibiotics to an animal, humans included, there is a number of consequences to their inner "microenvironment", affecting their microbiota and its susceptibility to antibiotics; these effects would not be discussed in this book. The intensive use of antibiotics and other biocide agents, within houses, hospitals and farms, affect the bacterial populations within those "microenvironments"; also, these effects would not be discussed here. To find antibiotics and/or antibiotic resistant organisms would be hardly surprising in such settings. However, antibiotics and/or antibiotic resistant organisms are commonly found both, in environments outside the bodies of organisms to which industrialized antibiotics have been administered; and in environments outside the walls of facilities where antibiotics are used intensively. These are going to be analyzed with detail in this book.

It is possible to coarsely divide the environment into two categories: urban and rural; these categories can be sub-divided in a number of ways, that will be briefly discussed below.

1.3.1 Urban environments

The definition of an "urban area" varies from country to country, and can be based on the total number of inhabitants, the population density (number of inhabitants per

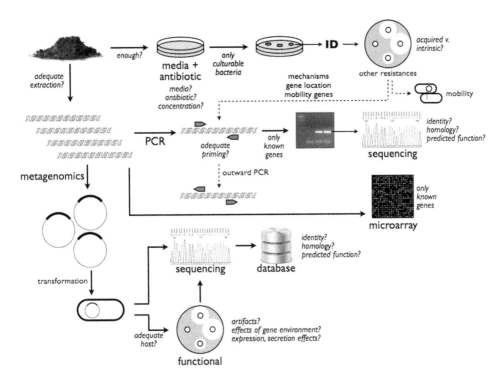

Figure 1.7 **Looking for resistant organisms or resistance genes.** An environmental sample (soil, water, feces) can be analyzed by culture- or molecular-based procedures. Culture-based methods (top) rely on our limited ability to grow bacteria in the lab; they are also limited by the size of the processed sample, the election of media, antibiotics and concentrations, and our actual interpretation of resistance. Mobility of resistance determinants, from the organism they were originally found in, to other strains, can be further analyzed when actually having a grown resistant isolate. Molecular methods were first developed based on PCR, which can be very sensitive, but also limited to known genes, and can have many causes of false positives and negatives; amplicons can be sequenced and then compared to known resistance genes. Some modifications allow for the description of the genetic entourage of the detected resistance gene; microarrays enable the detection of many different genes from the same sample, but with limitations similar to those of PCR. Metagenomics include the cloning of DNA fragments found in the environmental sample; these cloned fragments can directly undergo sequencing, and resulting sequences can be analyzed against resistance-gene databases so that known or putative resistance genes can be detected; or can be introduced into a new host cell (often *E. coli*) so that their expression can produce a detectable resistance phenotype. While capable of detecting real or possible resistance genes from the great majority of – unculturable – environmental bacteria, the ability of metagenomics to reveal actual threats other than well-known resistance genes remains to be proven.

area unit), or even the number of buildings per area unit. More than half of the human population, around 4 billion people, now lives in urban areas; small cities have a few hundred thousand inhabitants, and large cities have millions – or tens of millions. Additionally, there are cities with adequate sanitation services (*e.g.*, garbage and sewage disposal, pest control), minimal human or animal open defecation or urination, and effective environmental pollution controls, mainly in developed countries; and there are cities were wastewater runs in the open, open defecation from stray animals and homeless people is common, and air pollution can be several fold above recommended healthy limits, all typical of backward countries (a much more precise term than the euphemistic "developing" countries (Hobsbawm, 1994): at least in the lifespan of this author, no "developing" country have actually become developed; but lets leave it as "non-developed" for the sake of political correctness). There is limited information on the impact of each of these variables on the presence of antibiotics and antibiotic resistant bacteria in urban environments, although a great deal of information indicates that there could be important influence from such variables. Furthermore, there are some issues that must be kept in mind:

– In urban environments, antibiotics are mostly used therapeutically, on sick humans or pet animals. Most antibiotics excreted by such individuals end up in the sewage (although a little amount may be released by open defecation and urination). However, in countries where discarded drugs are dumped along with general garbage, and waste management is deficient or lacking, large amounts of antibiotics can be released into the urban environment. Little is known about this.
– Antibiotic-resistant bacteria infecting or colonizing humans or pet animals in urban settings, can easily spread outside of houses or hospitals, both along with infected or colonized humans and animals, or in their waste; however, little information is available on the presence of bacteria in urban dust, and in crowded urban environments (Rosas et al., 2011). Only very recently have metagenomic approaches yield some data about the microbiota of such urban settings as public transport systems, or even the urban fauna (Ehrenberg, 2015). Pathogens carrying antibiotic resistance genes were abundantly found in paper currency from India (Jalali et al., 2015), revealing one of the many unsuspected vehicles for resistance spread. In open environments, airborne bacteria are rare; the survival or pathogens in the soil (e.g., those released in urine or feces) depends on weather conditions, and on surface runoff collection. Potential interactions of such microorganisms among themselves and with chemical or physical agents in the urban environment, are mostly unknown.
– Several compounds that co-select for antibiotic resistance are typical of urban environments, including household or hospital disinfectants, heavy metals, and even air pollutants such as ozone. Again, the presence of such agents depends on the population density and the development level of such urban areas, among other things, and their influence upon resistance is mostly unknown.

In the end, the available evidence suggests that urban environments exert a pressure upon bacterial populations. In following chapters, this evidence and resulting working theories will be analyzed.

1.3.2 Rural environments

Outside of cities, the "environment" is of extreme diversity. However, the most relevant variable, for the purposes of this book, seems to be human influence. There are many ways in which the noxious components of urban environments "leak" into rural settings, the most obvious and studied being wastewater, but hardly the only one. Hence, a gradient can be established, in terms of simple distance from human settlements, or of specific means of "extending" human influence, such as along water streams, or even within aerial animals. Antibiotics and/or antibiotic resistant bacteria can be found across these gradients, establishing gradients of their own, which can help pinpoint their origin; non-antibiotic selective or maintenance pressures must also be considered along with other human influences. Of particular interest are two additional scenarios, that place copious amounts of antibiotics in rural environments: (1) the release of antibiotics from drug manufacturing factories (often in industrial areas but outside of cities), usually through wastewater dumped into water bodies; and (2) the agricultural use of antibiotics, to treat or prevent infections in animals or plants, or as "growth promoters"; this form of use, in addition to releasing antibiotics, is also selecting for resistant bacteria, that are released to the environment, sometimes in huge amounts, as when manure is used as fertilizer.

Antibiotic resistance can also be found in far away, pristine environments, indicating the local generation of resistance genes or phenotypes, independently of human influence. Although there are many papers reporting such findings, linkage to antibiotic-producing organisms and/or antibiotic presence in such environments, is rarely (or incompletely) made. While many authors find it perplexing to find resistance in such secluded settings, it is rather their perplexity which is perplexing: antibiotics and, of course, antibiotic resistance, have been around for millions of years.

Most papers on antibiotics and resistance in non-urban environments, either report their finding in assumedly pristine settings, or compare samples along a real or perceived gradient of human influence. All sorts of combinations can be found in such literature: samples of soil, water or wild animals' feces, analyzed by culture- and/or molecular-based methods. An analysis of these reports will occupy most of the remaining pages of this book. However, it is important to highlight two particular issues regarding these kinds of studies:

–   Intrinsic resistance is a major "contaminant" in reports of resistance in the environment. Research groups involved with this kind of research seem seldom to be multi-disciplinary, resulting in dismal errors both, in methodology and interpretation of results; peer-review seems to be equally affected. The same goes to metagenomic studies using resistance gene databases that include genes of questionable relevance to the subject.

–   There is a preoccupying lack of consensus regarding the very purpose of resistance surveillance in the environment. To some authors, finding a resistance gene in an unknown, non-culturable bacterial species in the soil or water of a remote location, indicates the clear risk of such determinant to eventually appear in a human pathogen in a clinical scenario. To others, it is a demonstration of the reach of human-related pollution of the environment, since high prevalence of resistance can only be explained by such determinants traveling from areas where antibiotic abuse is rife. Yet for others, it is a simple reminder that resistance has always been there. All three can be right or wrong, depending mostly on the definition of resistance.

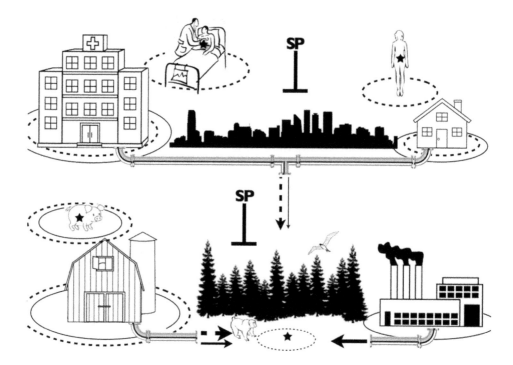

Figure 1.8 **Antibiotics, antibiotic resistance, and the environment; a graphic summary.** Antibiotics (stars) are used clinically, on inpatients and outpatients alike. While most of antibiotics excreted by patients end up in the sewage, some clinically-used antibiotics end up in the urban environment, especially in non-developed countries (continuous lines and circles). Antibiotics select for antibiotic resistant bacteria; such bacteria are shared by the patient to other people, and often escape the house or hospital, and are found in the urban environment and the sewage (dotted lines and circles). Carried by water (wastewater and runoff), resistant bacteria and some antibiotics can find their way to rural environments. Such environments also receive antibiotics and resistant bacteria from farm animals receiving antibiotics therapeutically or as food additives. Also, pharmaceutical factories can release antibiotics into these environments. A number of antibiotic-producing bacteria can add up to the ubiquitous presence of antibiotics in the environment; and such bacteria, and close neighbors, are also resistant. Resistant bacteria can be found in the soil, and in terrestrial and aerial wildlife. Finally, a number of known and unknown selective pressures (SP) act upon urban and non-urban environments, selecting and/or maintaining resistance determinants within bacterial populations.

–   The definition or assessment of the "pristine" quality of some environments must be carefully examined. Water streams, aerial animals, and even ecotourism, can bring resistant bacteria into such purportedly pristine environments.

REFERENCES

Aertsen, A., Van Houdt, R., Vanoirbeek, K. & Michiels, C. W. (2004) An SOS response induced by high pressure in *Escherichia coli*. *J. Bacteriol.*, 186, 6133–6141.

Amábile-Cuevas, C. F. & Arredondo-García, J. L. (2013) Nitrofurantoin, phenazopyridine, and the superoxide-response regulon *soxRS* of *Escherichia coli*. *J. Infect. Chemother.*, 19, 1135–1140.

Amábile-Cuevas, C. F. (1993) *Origin, evolution and spread of antibiotic resistance genes*, Austin, RG Landes.

Amábile-Cuevas, C. F. (2013) Antibiotic resistance: from Darwin to Lederberg to Keynes. *Microb. Drug Resist.*, 19, 73–87.

Amábile-Cuevas, C. F. & Chicurel, M. E. (1992) Bacterial plasmids and gene flux. *Cell*, 70, 189–199.

Amábile-Cuevas, C. F. & Chicurel, M. E. (1996) A possible role for plasmids in mediating the cell-cell proximity required for gene flux. *J. Theor. Biol.*, 181, 237–243.

Anzueto, A. & Norris, S. (2004) Clarithromycin in 2003: sustained efficacy and safety in an era of rising antibiotic resistance. *Int. J. Antimicrob. Agents*, 24, 1–17.

Arias, C. A., Panesso, D., Mcgrath, D. M., Qin, X., Mojica, M. F., Miller, C., Diaz, L., Tran, T. T., Rincon, S., Barbu, E. M., Reyes, J., Roh, J. H., Lobos, E., Sodergren, E., Pasqualini, R., Arap, W., Quinn, J. P., Shamoo, Y., Murray, B. E. & Weinstock, G. M. (2011) Genetic basis for *in vivo* daptomycin resistance in enterococci. *N. Engl. J. Med.*, 365, 892–900.

Arredondo-García, J. L. & Amábile-Cuevas, C. F. (2008) High resistance prevalence towards ampicillin, co-trimoxazole and ciprofloxacin, among uropathogenic *Escherichia coli* isolates in Mexico City. *J. Infect. Developing Countries*, 2, 350–353.

Babic, A., Berkmen, M. B., Lee, C. A. & Grossman, A. D. (2011) Efficient gene transfer in bacterial cell chains. *mBio*, 2, e00027-11.

Baharoglu, Z., Bikard, D. & Mazel, D. (2010) Conjugative DNA transfer induces the bacterial SOS response and promotes antibiotic resistance development through integron activation. *PLoS Genet.*, 6, e1001165.

Balsalobre, L., Ferrándiz, M. J., Liñares, J., Tubau, F. & De La Campa, A. G. (2003) Viridans group streptococci are donors in horizontal transfer of topoisomerase IV genes to *Streptococcus pneumoniae*. *Antimicrob. Agents Chemother.*, 47, 2072–2081.

Baquero, F. (2011) The 2010 Garrod Lecture: the dimensions of evolution in antibiotic resistance: *ex unibus plurum et ex pluribus unum*. *J. Antimicrob. Chemother.*, 66, 1659–1672.

Bayer, A. S., Schneider, T. & Sahl, H. G. (2013) Mechanisms of daptomycin resistance in *Staphylococcus aureus*: role of the cell membrane and cell wall. *Ann. NY Acad. Sci.*, 1277, 139–158.

Beaber, J. W., Hochhut, B. & Waldor, M. K. (2004) SOS response promotes horizontal dissemination of antibiotic resistance genes. *Nature*, 427, 72–74.

Benveniste, R. & Davies, J. (1973) Aminoglycoside antibiotic-inactivating enzymes in actinomycetes similar to those present in clinical isolates of antibiotic-resistant bacteria. *Proc. Natl. Acad. Sci. USA*, 70, 2276–2280.

Bergh, O., Børsheim, K. Y., Bratbak, G. & Heldal, M. (1989) High abundance of viruses found in aquatic environments. *Nature*, 340, 467–468.

Breeze, A. S. & Obaseiki-Ebor, E. E. (1983) Transferable nitrofuran resistance conferred by R-plasmids in clinical isolates of *Escherichia coli*. *J. Antimicrob. Chemother.*, 12, 459–467.

Brenwald, N. P., Appelbaum, P., Davies, T. & Gill, M. J. (2002) Evidence for efflux pumps, other than PmrA, associated with fluoroquinolone resistance in *Streptococcus pneumoniae*. *Clin. Microbiol. Infect.*, 9, 140–143.

Brown, E. W., Leclerc, J. E., Li, B., Payne, W. L. & Cebula, T. S. (2001) Phylogenetic evidence for horizontal transfer of *mutS* alleles among naturally occurring *Escherichia coli* strains. *J. Bacteriol.*, 183, 1631–1644.

Bushman, F. (2002) *Lateral DNA transfer*, Cold Spring Harbor, Cold Spring Harbor Laboratory Press.

Campanile, F., Mongelli, G., Bongiorno, D., Adembri, C., Ballardini, M., Falcone, M., Menichetti, F., Repetto, A., Sabia, C., Sartor, A., Scarparo, C., Tascini, C., Venditti, M. F. Z. & Stefani, S. (2013) Worrisome trend of new multiple mechanisms of linezolid resistance in staphylococcal clones diffused in Italy. *J. Clin. Microbiol.*, 51, 1256–1259.

Ceccherini, M. T., Poté, J., Kay, E., Van, V. T., Maréchal, J., Pietramellara, G., Nannipieri, P., Vogel, T. M. & Simonet, P. (2003) Degradation and transformability of DNA from transgenic leaves. *Appl. Environ. Microbiol.*, 69, 673–678.

Charpentier, X., Kay, E., Schneider, D. & Shuman, H. A. (2011) Antibiotics and UV radiation induce competence for natural transformation in *Legionella pneumophila*. *J. Bacteriol.*, 193, 1114–1121.

Chiura, H. X., Kogure, K., Hagemann, S., Ellinger, A. & Velimirov, B. (2011) Evidence for particle-induced horizontal gene transfer and serial transduction between bacteria. *FEMS Microbiol. Ecol.*, 76, 576–591.

Choi, I. G. & Kim, S. H. (2007) Global extent of horizontal gene transfer. *Proc. Natl. Acad. Sci. USA*, 104, 4489–4494.

Ciofu, O., Beveridge, T. J., Kadurugamuwa, J., Walther-Rasmussen, J. & Hoiby, N. (2000) Chromosomal beta-lactamase is packaged into membrane vesicles and secreted from *Pseudomonas aeruginosa*. *J. Antimicrob. Chemother.*, 45, 9–13.

Cirz, R. T., Chin, J. K., Andes, D. R., De Crécy-Lagard, V., Craig, W. A. & Romesberg, F. E. (2005) Inhibition of mutation and combating the evolution of antibiotic resistance. *PLoS Biol.*, 3, e176.

Clemente, J. C., Pehrsson, E. C., Blaser, M. J., Sandhu, K., Gao, Z., Wang, B., Magris, M., Hidalgo, G., Contreras, M., Noya-Alarcón, Ó., Lander, O., Mcdonald, J., Cox, M., Walter, J., Oh, P. L., Ruiz, J. F., Rodriguez, S., Shen, N., Song, S. J., Metcalf, J., Knight, R., Dantas, G. & Dominguez-Bello, M. G. (2015) The microbiome of uncontacted Amerindians. *Sci. Adv.*, 1, e1500183.

Cohen, S. N. (1976) Transposable genetic elements and plasmid evolution. *Nature*, 263, 731–738.

Colomer-Lluch, M., Imamovic, L., Jofre, J. & Muniesa, M. (2011) Bacteriophages carrying antibiotic resistance genes in fecal waste from cattle, pigs, and poultry. *Antimicrob. Agents Chemother.*, 55, 4908–4911.

Cooper, T. F. & Heinemann, J. A. (2000) Transfer of conjugative plasmids and bacteriophage lambda occurs in the presence of antibiotics that prevent de novo gene expression. *Plasmid*, 43, 171–175.

Courvalin, P. (1990) Plasmid-mediated 4-quinolone resistance: a real or apparent absence? *Antimicrob. Agents Chemother.*, 34, 681–684.

Courvalin, P. (2008) Predictable and unpredictable evolution of antibiotic resistance. *J. Intern. Med.*, 264, 4–16.

Coyne, S., Courvalin, P. & Périchon, B. (2011) Efflux-mediated antibiotic resistance in *Acinetobacter* spp. *Antimicrob. Agents Chemother.*, 55, 947–953.

Dagan, T. & Martin, W. (2009) Getting a better picture of microbial evolution en route to a network of genomes. *Phil. Trans. R. Soc. B*, 364, 2187–2196.

Davies, J. (1990) What are antibiotics? Archaic functions for modern activities. *Mol. Microbiol.*, 4, 1227–1232.

Davin-Regli, A. & Pagès, J.-M. (2007) Regulation of efflux pumps in Enterobacteriaceae: genetic and chemical effectors. In Amábile-Cuevas, C. F. (Ed.) *Antimicrobial resistance in bacteria*. Wymondham, UK, Horizon Bioscience.

De Vries, J., Meier, P. & Wackernagel, W. (2001) The natural transformation of the soil bacteria *Pseudomonas stutzeri* and *Acinetobacter* sp. by transgenic plant DNA strictly depends on homologous sequences in the recipient cells. *FEMS Microbiol. Lett.*, 195, 211–215.

Delissalde, F. & Amábile-Cuevas, C. F. (2004) Comparison of antibiotic susceptibility and plasmid content, between biofilm producing and non-producing clinical isolates of *Pseudomonas aeruginosa*. *Int. J. Antimicrob. Agents*, 24, 405–408.

Demple, B. & Amábile-Cuevas, C. F. (2003) Multiple resistance mediated by individual genetic loci. In Amábile-Cuevas, C. F. (Ed.) *Multiple drug resistant bacteria*. Wymondham, Horizon Scientific Press.

Díaz-Mejía, J. J., Amábile-Cuevas, C. F., Rosas, I. & Souza, V. (2008) An analysis of the evolutionary relationships of integron integrases, with emphasis on the prevalence of class 1 integron in *Escherichia coli* isolates from clinical and environmental origins. *Microbiol.*, 154, 94–102.

Dionisio, F., Matic, I., Radman, M., Rodrigues, O. R. & Taddei, F. (2002) Plasmids spread very fast in heterogeneous bacterial communities. *Genetics*, 162, 1525–1532.

Duffin, P. M. & Seifert, H. S. (2012) Genetic transformation of *Neisseria gonorrhoeae* shows strand preference. *FEMS Microbiol. Lett.*, 334, 44–48.

Ehrenberg, R. (2015) Urban microbes unveiled. *Nature*, 522, 399–400.

Ferguson, G. C., Heinemann, J. A. & Kennedy, M. A. (2002) Gene transfer between *Salmonella enterica* serovar Typhimurium inside epithelial cells. *J. Bacteriol.*, 184, 2235–2242.

Ferrándiz, M. J., Fenoll, A., Liñares, J. & De La Campa, A. (2000) Horizontal transfer of *parC* and *gyrA* in fluoroquinolone-resistant clinical isolates of *Streptococcus pneumoniae*. *Antimicrob. Agents Chemother.*, 44, 840–847.

Firth, N. (2003) Evolution of antimicrobial multi-resistance in gram-positive bacteria. In Amábile-Cuevas, C. F. (Ed.) *Multiple drug resistant bacteria*. Wymondham, Horizon Scientific Press.

Flynn, K. J. & Swanson, M. S. (2014) Integrative conjugative element ICE-b ox confers oxidative stress resistance to *Legionella pneumophila in vitro* and in macrophages. *mBio*, 5, e01091-14.

Foster, T. J. (1983) Plasmid-determined resistance to antimicrobial drugs and toxic metal ions in bacteria. *Microbiol. Rev.*, 47, 361–409.

Fouhy, F., Stanton, C., Cotter, P. D., Hill, C. & Walsh, F. (2015) Proteomics as the final step in the functional metagenomis study of antimicrobial resistance. *Front. Microbiol.*, 6, 172.

Fraimow, H. S. (2003) Vancomycin resistant enterococci and methicillin resistant *Staphylococcus aureus*. In Amábile-Cuevas, C. F. (Ed.) *Multiple drug resistant bacteria*. Wymondham, Horizon Scientific Press.

Fuchs, L. Y., Reyna, F., Chihu, L. & Carrillo, B. (1996) Molecular aspects of fluoroquinolone resistance. In Amábile-Cuevas, C. F. (Ed.) *Antibiotic resistance: from molecular basics to therapeutic options*. Austin/New York, Landes/Chapman & Hall.

Fuentes, A. M. & Amábile Cuevas, C. F. (1997) Mercury induces multiple antibiotic resistance in *Escherichia coli* through activation of SoxR, a redox-sensing regulatory protein. *FEMS Microbiol. Lett.*, 154, 385–388.

Fuentes, A. M., Díaz-Mejía, J. J., Maldonado-Rodríguez, R. & Amábile Cuevas, C. F. (2001) Differential activities of the SoxR protein of *Escherichia coli*: SoxS is not required for gene activation under iron deprivation. *FEMS Microbiol. Lett.*, 201, 271–275.

Gerdes, K., Boe, L., Andersson, P. & Molin, S. (1986) Plasmid stabilization in populations of bacterial cells. In Levy, S. B. & Novick, R. P. (Eds.) *Antibiotic resistance genes: ecology, transfer, and expression*. Cold Spring Harbor, Cold Spring Harbor Laboratory.

Ghigo, J. M. (2001) Natural conjugative plasmids induce bacterial biofilm development. *Nature*, 412, 442–445.

Gilbert, P. & McBain, A. (2003) Potential impact of increased use of biocides in consumer products on prevalence of antibiotic resistance. *Clin. Microbiol. Rev.*, 16, 189–208.

Gilbert, P., McBain, A. & Lindsay, S. (2007) Biofilms, multi-resistance, and persistence. In Amábile-Cuevas, C. F. (Ed.) *Antimicrobial resistance in bacteria*. Wymondham, UK, Horizon Bioscience.

Girgis, H. S., Hottes, A. K. & Tavazoie, S. (2009) Genetic architecture of intrinsic antibiotic susceptibility. *PLoS One*, 4, e5629.

Guerin, E., Cambray, G., Sanchez-Alberola, N., Campoy, S., Erill, I., Da Re, S., Gonzalez-Zorn, B., Barbé, J., Ploy, M. C. & Mazel, D. (2009) The SOS response controls integron recombination. *Science*, 324, 1034.

Guerra, B., Soto, S., Helmuth, R. & Mendoza, M. C. (2002) Characterization of a self-transferable plasmid from *Salmonella enterica* serotype Typhimurium clinical isolates carrying two integron-borne gene cassettes together with virulence and drug resistance genes. *Antimicrob. Agents Chemother.*, 46, 2977–2981.

Guglielmini, J., Quintais, L., Garcillán-Barcia, M. P., de la Cruz, F. & Rocha, E. P. C. (2011) The repertoire of ICE in prokaryotes underscores the unity, diversity, and ubiquity of conjugation. *PLoS Genet*, 7, e1002222.

Hamilton, H. L. & Dillard, J. P. (2006) Natural transformation of *Neisseria gonorrhoeae*: from DNA donation to homologous recombination. *Mol. Microbiol.*, 59, 376–385.

Hansen, L. H., Johannesen, E., Burmølle, M., Sørensen, A. H. & Sørensen, S. J. (2004) Plasmid-encoded multidrug efflux pump conferring resistance to olaquindox in *Escherichia coli*. *Antimicrob. Agents Chemother.*, 48, 3332–3337.

Hansen, S. K., Rainey, P. B., Haagesen, J. A. J. & Molin, S. (2007) Evolution of species interactions in a biofilm community. *Nature*, 445, 533–536.

Heinemann, J. A. & Ankenbauer, R. G. (1993a) Retrotransfer in *Escherichia coli* conjugation: bidirectional exchange or *de novo* mating? *J. Bacteriol.*, 175, 583–588.

Heinemann, J. A. & Ankenbauer, R. G. (1993b) Retrotransfer of IncP plasmid R751 from *Escherichia coli* maxicells: evidence for the genetic sufficiency of self-transferable plasmids for bacterial conjugation. *Mol. Microbiol.*, 10, 57–62.

Heinemann, J. A., Ankenbauer, R. G. & Amábile-Cuevas, C. F. (2000) Do antibiotics maintain antibiotic resistance? *Drug Discov. Today*, 5, 195–204.

Heinemann, J. A. & Silby, M. W. (2003) Horizontal gene transfer and the selection of antibiotic resistance. In Amábile-Cuevas, C. F. (Ed.) *Multiple drug resistant bacteria*. Wymondham, Horizon.

Hobsbawm, E. (1994) *The age of extremes*, New York, Vintage Books.

Hoffman, L. R., D'Argenio, D. A., Maccoss, M. J., Zhang, Z., Jones, R. A. & Miller, S. I. (2005) Aminoglycoside antibiotics induce bacterial biofilm formation. *Nature*, 436, 1171–1175.

Huovinen, P., Sundström, L., Swedberg, G. & Sköld, O. (1995) Trimethoprim and sulfonamide resistance. *Antimicrob. Agents Chemother.*, 39, 279–289.

Itzek, A., Zheng, L., Chen, Z., Merritt, J. & Kreth, J. (2011) Hydrogen peroxide-dependent DNA release and transfer of antibiotic resistance genes in *Streptococcus gordonii*. *J. Bacteriol.*, 193, 6912–6922.

Jacoby, G. A. (2009) AmpC β-lactamases. *Clin. Microbiol. Rev.*, 22, 161–182.

Jain, R., Rivera, M. C. & Lake, J. A. (1999) Horizontal gene transfer among genomes: the complexity hypothesis. *Proc. Natl. Acad. Sci. USA*, 96, 3801–3806.

Jalali, S., Kohli, S., Latka, C., Bhatia, S., Vellarikal, S. K., Sivasubbu, S., Scaria, V. & Ramachandran, S. (2015) Screening currency notes for microbial pathogens and antibiotic resistance genes using a shotgun metagenomic approach. *PLoS One*, 10, e0128711.

Jiménez-Arribas, G., Léautaud, V. & Amábile-Cuevas, C. F. (2001) Regulatory locus *soxRS* partially protects *Escherichia coli* against ozone. *FEMS Microbiol Letters*, 195, 175–177.

Kittler, S., Fischer, S., Abdulmawjood, A., Glünder, G. & Klein, G. (2013) Effect of bacteriophage application on *Campylobacter jejuni* loads in commercial broiler flocks. *Appl. Environ. Microbiol.*, 23, 7525–7535.

Kohanski, M., Dwyer, D., Hayete, B., Lawrence, C. & Collins, J. (2007) A common mechanism of cellular death induced by bactericidal antibiotics. *Cell*, 130, 797–810.

Koonin, E. V. & Wolf, Y. I. (2012) Evolution of microbes and viruses: a paradigm shift in evolutionary biology? *Front. Cell. Inf. Microbio.*, 2, 119.

Kuan, C. T. & Tessman, I. (1991) LexA protein of *Escherichia coli* represses expression of the Tn*5* transposase gene. *J. Bacteriol.*, 173, 6406–6410.

Kurenbach, B., Marjoshi, D., Amábile Cuevas, C. F., Ferguson, G. C., Godsoe, W., Gibson, P. & Heinemann, J. A. (2015) Sublethal exposure to commercial formulations of the herbicides dicamba, 2,4-dichlorophenoxyacetic acid, and glyphosate cause changes in antibiotic susceptibility in *Escherichia coli* and *Salmonella enterica* serovar Typhimurium. *mBio*, 6, e00009-15.

Lau, S. K. P., Woo, P. C. Y., To, A. P. C., Lau, A. T. K. & Yuen, K. Y. (2004) Lack of evidence that DNA in antibiotic preparations is a source of antibiotic resistance genes in bacteria from animal or human sources. *Antimicrob. Agents Chemother.*, 48, 3141–3146.

Laxminarayan, R., Van Boeckel, T. & Teillant, A. (2015) The economic costs of withdrawing antimicrobial growth promoters from the livestock sector. *OECD Food, Agriculture and Fisheries Papers*, 78.

Leclercq, S., Gilbert, C. & Cordaux, R. (2012) Cargo capacity of phages and plasmids and other factors influencing horizontal transfers of prokaryote transposable elements. *Mob. Genet. Elements*, 2, 115–118.

Léon, G. & Roy, P. H. (2009) Potential role of group IIC-attC introns in integron cassette formation. *J. Bacteriol.*, 191, 6040–6051.

Levin, B. (2004) Noninherited resistance to antibiotics. *Science*, 305, 1578–1579.

Levy, S. B. (2002) Active efflux, a common mechanism for biocide and antibiotic resistance. *J. Appl. Microbiol.*, 92 (suppl.), 65S–71S.

Li, Y. H., Lau, P. C. Y., Lee, J. H., Ellen, R. P. & Cvitkovitch, D. G. (2001) Natural genetic transformation of *Streptococcus mutans* growing in biofilms. *J. Bacteriol.*, 183, 897–908.

Linares, J. F., Gustafsson, I., Baquero, F. & Martinez, J. L. (2006) Antibiotics as intermicrobial signaling agents instead of weapons. *Proc. Natl. Acad. Sci. USA*, 103, 19484–19489.

Lister, P. D., Wolter, D. J. & Hanson, N. D. (2009) Antibacterial-resistant *Pseudomonas aeruginosa*: clinical impact and complex regulation of chromosomally encoded resistance mechanisms. *Clin. Microbiol. Rev.*, 22, 582–610.

Lu, K., Asano, R. & Davies, J. (2004) Antimicrobial resistance gene delivery in animal feeds. *Emerg. Infect. Dis.*, 10, 679–683.

Mahoney, T. F. & Silhavy, T. J. (2013) The Cpx stress response confers resistance to some, but not all, bactericidal antibiotics. *J. Bacteriol.*, 195, 1869–1874.

Martínez, J. L. & Baquero, F. (2000) Mutation frequencies and antibiotic resistance. *Antimicrob. Agents Chemother.*, 44, 1771–1777.

Martínez, J. L., Coque, T. M. & Baquero, F. (2015) What is a resistance gene? Ranking risk in resistomes. *Nat. Rev. Microbiol.*, 13, 116–123.

Martínez-Puchol, S., Gomes, C., Pons, M. J., Ruiz-Roldán, L., Torrents De La Peña, A., Ochoa, T. J. & Ruiz, J. (2015) Development and analysis of furazolidone-resistant *Escherichia coli* mutants. *Acta Pathol. Microbiol. Immunol. Scand.*

McArthur, A. G., Waglechner, N., Nizam, F., Yan, A., Azad, M. A., Baylay, A. J., Bhullar, K., Canova, M. J., De Pascale, G., Ejim, L., Kalan, L., King, A. M., Koteva, K., Morar, M., Mulvey, M. R., O'brien, J. S., Pawlowski, A. C., Piddock, L. J. V., Spanogiannopoulos, P., Sutherland, A. D., Tang, I., Taylor, P. L., Thaker, M., Wang, W., Yan, M., Yu, T. & Wright, G. D. (2013) The comprehensive antibiotic resistance database. *Antimicrob. Agents Chemother.*, 57, 3348–3357.

Meaden, S. & Koskella, B. (2013) Exploring the risks of phage application in the environment. *Front. Microbiol.*, 4, 358.

Medrano-Soto, A., Moreno-Hagelsieb, G., Vinuesa, P., Christen, J. A. & Collado-Vides, J. (2004) Successful lateral transfer requires codon usage compatibility between foreign genes and recipient genomes. *Mol. Biol. Evol.*, 21, 1884–1894.

Miller, C., Thomsen, L. E., Gaggero, C., Mosseri, R., Ingmer, H. & Cohen, S. N. (2004) SOS response induction by β-lactams and bacterial defense against antibiotic lethality. *Science*, 305, 1629–1631.

Montero, C. I., Stock, F. & Murray, P. R. (2008) Mechanisms of resistance to daptomycin in *Enterococcus faecium*. *Antimicrob. Agents Chemother.*, 52, 1167–1170.

Moritz, E. M. & Hergenrother, P. J. (2007) The prevalence of plasmids and other mobile genetic elements in clinically important drug-resistance bacteria. In Amábile-Cuevas, C. F. (Ed.) *Antimicrobial resistance in bacteria*. Wymondham, Horizon Bioscience.

Naas, T., Mikami, Y., Imai, T., Poirel, L. & Nordmann, P. (2001) Characterization of In53, a class 1 plasmid- and composite transposon-located integron of *Escherichia coli* which carries an unusual array of gene cassettes. *J. Bacteriol.*, 183, 235–249.

Nijland, R., Burgess, J. G., Errington, J. & Veening, J. W. (2010) Transformation of environmental *Bacillus subtilis* isolates by transiently inducing genetic competence. *PLoS One*, 5, e9724.

Nogueira, T., Rankin, D. J., Touchon, M., Taddei, F., Brown, S. P. & Rocha, E. P. C. (2009) Horizontal gene transfer of the secretome drives the evolution of bacterial cooperation and virulence. *Curr. Biol.*, 19, 1683–1691.

Nunoshiba, T., deRojas-Walker, T., Wishnok, J. S., Tannenbaum, S. R. & Demple, B. (1993) Activation by nitric oxide of an oxidative-stress response that defends *Escherichia coli* against activated macrophages. *Proc. Natl. Acad. Sci. USA*, 90, 9993–9997.

Ochman, H., Lawrence, J. G. & Groisman, E. A. (2000) Lateral gene transfer and the nature of bacterial innovation. *Nature*, 405, 299–304.

Olaitan, A. O., Morand, S. & Rolain, J. M. (2014) Mechanisms of polymyxin resistance: acquired and intrinsic resistance in bacteria. *Front. Microbiol.*, 5, 643.

Oren, Y., Smith, M. B., Johns, N. I., Zeevi, M. K., Biran, D., Ron, E. Z., Corander, J., Wang, H. H., Alm, E. J. & Pupko, T. (2014) Transfer of noncoding DNA drives regulatory rewiring in bacteria. *Proc. Natl. Acad. Sci. USA*, 111, 16112–16117.

Pontiroli, A., Rizzi, A., Simonet, P., Daffonchio, D., Vogel, T. M. & Monier, J. M. (2009) Visual evidence of horizontal gene transfer between plants and bacteria in the phytosphere of transplastomic tobacco. *Appl. Environ. Microbiol.*, 75, 3314–3322.

Pruden, A. & Arabi, M. (2012) Quantifying anthropogenic impacts on environmental reservoirs of antibiotic resistance. In Keen, P. L. & Montforts, M. H. M. M. (Eds.) *Antimicrobial resistance in the environment*. New Jersey, John Wiley & Sons.

Rankin, D. J., Rocha, E. P. C. & Brown, S. P. (2010) What traits are carried on mobile genetic elements, and why? *Heredity*.

Raze, D., Dardenne, O., Hallut, S., Martinez-Bueno, M., Coyette, J. & Ghuysen, J. M. (1998) The gene encoding the low-affinity penicillin-binding protein 3r in *Enterococcus hirae* S185R is borne on a plasmid carrying other antibiotic resistance determinants. *Antimicrob. Agents Chemother.*, 42, 543–549.

Renggli, S., Keck, W., Jenal, U. & Ritz, D. (2013) Role of autofluorescence in flow cytometric analysis of *Escherichia coli* treated with bactericidal antibiotics. *J. Bacteriol.*, 195, 4067–4073.

Reyes, A., Haynes, M., Hanson, N., Angly, F. E., Heath, A. C., Rohwer, F. & Gordon, J. I. (2010) Viruses in the faecal microbiota of monozygotic twins and their mothers. *Nature*, 466, 334–338.

Rezzonico, F., Stockwell, V. O. & Duffy, B. (2009) Plant agricultural streptomycin formulations do not carry antibiotic resistance genes. *Antimicrob. Agents Chemother.*, 53, 3173–3177.

Rosas, I., Amábile-Cuevas, C. F., Calva, E. & Osornio-Vargas, A. (2011) Animal and human waste as components of urban dust pollution: health implications. In Nriagu, J. O. (Ed.) *Encyclopedia of environmental health*. Amsterdam, Elsevier.

Rumbo, C., Fernández-Moreira, E., Merino, M., Poza, M., Mendez, J. A., Soares, N. C., Mosquera, A., Chaves, F. & Bou, G. (2011) Horizontal transfer of the OXA-24 carbapenemase gene via outer membrane vesicles: a new mechanism of dissemination of carbapenem resistance genes in *Acinetobacter baumannii*. *Antimicrob. Agents Chemother.*, 55, 3084–3090.

Sandegren, L., Lindqvist, A., Kahlmeter, G. & Andersson, D. I. (2008) Nitrofurantoin resistance mechanisms and fitness cost. *J. Antimicrob. Chemother.*, 62, 495–503.

Sedgwick, S. G., Thomas, S. M., Hughes, V. M., Lodwick, D. & Strike, P. (1989) Mutagenic DNA repair genes on plasmids from the 'pre-antibiotic era'. *Mol. Gen. Genet.*, 218, 323–329.

Sengupta, S., Chattopadhyay, M. K. & Grossart, H. P. (2013) The multifaceted roles of antibiotics and antibiotic resistance in nature. *Front. Microbiol.*, 4, 47.

Skippington, E. & Ragan, M. A. (2012) Phlogeny rather than ecology or lifestyle biases the construction of *Escherichia coli-Shigella* genetic exchange communities. *Open Biol.*, 2, 120112.

Smillie, C. S., Smith, M. B., Friedman, J., Cordero, O. X., David, L. A. & Alm, E. J. (2011) Ecology drives a global network of gene exchange connecting the human microbiome. *Nature*, 480, 241–244.

Soares, G. M. S., Figueiredo, L. C., Faveri, M., Cortelli, S. C., Duarte, P. M. & Feres, M. (2012) Mechanisms of action of systemic antibiotics used in periodontal treatment and mechanisms of bacterial resistance to these drugs. *J. Appl. Oral Sci.*, 20, 295–309.

Strahilevitz, J., Jacoby, G. A., Hooper, D. C. & Robicsek, A. (2009) Plasmid-mediated quinolone resistance: a multifaceted threat. *Clin. Microbiol. Rev.*, 22, 664–689.

Sulakvelidze, A., Alavidze, Z. & Morris Jr., J. G. (2001) Bacteriophage therapy. *Antimicrob. Agents Chemother.*, 45, 649–659.

Sunde, M. (2005) Class I integron with a group II intron detected in an *Escherichia coli* strain from a free-range reindeer. *Antimicrob. Agents Chemother.*, 49, 2512–2514.

Tetz, G. V., Artemenko, N. K. & Tetz, V. V. (2009) Effect of DNase and antibiotics on biofilm characteristics. *Antimicrob. Agents Chemother.*, 53, 1204–1209.

Tewhey, R., Gu, B., Kelesidis, T., Charlton, C., Bobenchik, A., Hindler, J., Schork, N. J. & Humphries, R. M. (2014) Mechanisms of linezolid resistance among coagulase-negative staphylococci determined by whole-genome sequencing. *mBio*, 5, e00894-14.

Van Bambeke, F., Glupczynsky, Y., Plésiat, P., Pechère, J. C. & Tulkens, P. M. (2003) Antibiotic efflux pumps in prokaryotic cells: occurrence, impact on resistance and strategies for the future of antimicrobial therapy. *J Antimicrob. Chemother.*, 51, 1055–1065.

Van Hoek, A. H. A. M., Mevius, D., Guerra, B., Mullany, P., Roberts, A. P. & Aarts, H. J. M. (2011) Acquired antibiotic resistance genes: an overview. *Front. Microbiol.*, 2, 203.

Vesić, D. & Kristich, C. J. (2012) MurAA is required for intrinsic cephalosporin resistance of *Enterococcus faecalis*. *Antimicrob. Agents Chemother.*, 56, 2443–2451.

Wang, Q., Sun, F. J., Liu, Y., Xiong, L. R., Xie, L. L. & Xia, P. Y. (2010) Enhancement of biofilm formation by subinhibitory concentrations of macrolides in icaADBC-positive and -negative clinical isolates of *Staphylococcus epidermidis*. *Antimicrob. Agents Chemother.*, 54, 2707–2711.

Wang, Y., Lv, Y., Cai, J., Schwarz, S., Cui, L., Hu, Z., Zhang, R., Li, J., Zhao, Q., He, T., Wang, D., Wang, Z., Shen, Y., Li, Y., Feßler, A. T., Wu, C., Yu, H., Deng, X., Xia, X. & Shen, J. (2015) A novel gene, *optrA*, that confers transferable resistance to oxazolidinones and phenicols and its presence in *Enterococcus faecalis* and *Enterococcus faecium* of human and animal origin. *J. Antimicrob. Chemother.*, 70, 2182–2190.

Webb, V. & Davies, J. (1993) Antibiotic preparations contain DNA: a source of drug resistance genes? *Antimicrob. Agents Chemother.*, 37, 2379–2384.

Wolkow, C. A., Deboy, R. T. & Craig, N. L. (1996) Conjugating plasmids are preferred targets for Tn7. *Genes Develop.*, 10, 2145–2157.

Wozniak, D. J. & Keyser, R. (2004) Effects of subinhibitory concentrations of macrolide antibiotics on *Pseudomonas aeruginosa. Chest,* 125, 62–69.

Xue, H., Cordero, O. X., Camas, F. M., Trimble, W., Meyer, F., Guglielmini, J., Rocha, E. P. C. & Polz, M. F. (2015) Eco-evolutionary dynamics of episomes among ecologically cohesive bacterial populations. *mBio,* 6, e00552-15.

Chapter 2

Have antibiotics and antibiotic resistance genes always been out there?

Naturally-occurring antibiotics, which is to say most of clinically-used antibacterial drugs, are old molecules. While the actual "role" these molecules play in microbial physiology and ecology is still controversial (see section 1.1.2), antibiotic biosynthetic pathways are millions of years old (unless, of course, The Earth is actually only 6,000 years old). J. Davies even propose that some antibiotics can be prebiotic molecules, which would then be *billions* of years old. However, antibiotic production is not a very common feature amongst microbes: nearly 50% of known antibiotics are produced by a single bacterial genus, *Streptomyces*, a tiny fraction of all bacterial diversity. Furthermore, although about 1% of soil actinomycetes produce streptomycin, which seems to be pretty common, the number of strains that synthesize other antibiotics is much smaller – in the 0.000001% order (Laskaris et al., 2012). Of course, the key word in the two previous sentences is "known": there could be many more antibiotics unknown to us, produced by as many unknown bacteria (or even known antibiotics produced by other, unknown bacteria – at risk of sounding Rumsfeld-ish). This was very elegantly demonstrated with the discovery of teixobactin: by growing previously uncultured soil bacteria in an ingenious device (iChip), an unknown species named *Eleftheria terrae*, related to *Aquabacteria*, not previously known to produce antibiotics, was discovered, along with a new peptidoglycan-synthesis inhibitor (Ling et al., 2015). (This important discovery was however announced with the preposterous claim that resistance was unlikely to arise; while some immediately disagreed, providing evidence (Hochberg and Jansen, 2015), a simple movie quotation perhaps should suffice: "life finds a way"). A relevant question remains: how much and how many naturally-produced antibiotics exist in the environment?

2.1 NATURALLY-OCCURRING ANTIBIOTICS (AND THEIR RESPECTIVE RESISTANCE GENES)

Even though a significant amount of soil bacteria can produce streptomycin, there are no reports of the detection of such antibiotic in pristine soils: the amount of streptomycin – and other antibiotics, released to the soil by producing bacteria seems to be below our detection capabilities. This is particularly relevant because: (a) streptomycin is a particularly stable molecule; (b) the detection range of current technology is in the order of micrograms or even nanograms per liter or kilogram (*e.g.*, 60 nM, or ~34 µg/L, using chromatography under ideal conditions), which clearly contrasts with (c) the minimal inhibitory concentration of streptomycin upon clinically-relevant bacteria, which is in the order of 4–64 μg/mL, *i.e.*, 100–1,000-fold higher. Although "the absence of evidence is not evidence of absence", it is possible to conclude that natural antibiotics are present in the soil, where all known producing bacteria have been obtained from, at concentrations well below those necessary to inhibit the growth of other microbes. It is also plausible that, should naturally-produced streptomycin actually exerts a selective pressure upon other, non-producing soil bacteria, such an scenario is restricted to the very close neighborhood of producing streptomycetes. The selective pressure posed by other, scarcer antibiotics in natural environments would therefore be confined to much narrower spaces, if at all. In other words, resistance mechanisms providing protection against very high concentrations of antibiotics, would only benefit bacteria living within micrometers or millimeters from antibiotic producers which are, in turn, rather rare. This notion has been acknowledged before as a limitation of the hypothesis pointing at the environment as a source of resistance (Baquero et al., 1998).

On the other hand, it is perhaps relevant to remember that the concept of pathogenic bacteria "do not surviving long in the soil" probably because "soil-inhabiting microbes, antagonistic to the pathogens . . . bring about their rapid destruction in the soil", was behind the very discovery of streptomycin (Waksman and Woodruff, 1940). If clinically-used antibiotics are not to be found in the soil, especially not at inhibitory concentrations, which are the substances produced by soil saprophytes that "not only inhibit the growth of pathogens, but bring about their destruction"? One likely candidate is a metabolite of *Bacillus pyocaneus* (today known as *P. aeruginosa*), called pyocyanin, discovered by Freudenreich in 1888, but very toxic and unstable to be developed for clinical use; the genus *Pseudomonas* produces many toxic compounds, but only pseudomonic acid, or mupirocin, has reach the clinic as a topic antibiotic against gram-positives. Tyrothricin, a mixture of gramicidin and tyrocidine, discovered by Dubos in 1939, is produced by another soil bacterium, *Bacillus brevis*. Perhaps these compounds do play a chemical warfare role in microbial ecology and are abundant enough to "antagonize pathogens" when they fall to the soil.

2.1.1   Known resistance genes from known antibiotic producers

The list of clinically-used, natural antibiotics, along with the producing organisms, is in section 1.1.1; it includes eleven drug families, with families mostly formed by semi-synthetic derivatives of a single original antibiotic molecule (except for most aminoglycosides and many beta-lactams). Therefore, there are much fewer than a hundred known bacterial species that produce antibiotics further developed for pharmaceutical purposes. Although it is assumed that antibiotic-producing bacteria also carry resistance genes in order to "avoid suicide", there is limited information on the actual nature of such genes, except for the most common, old antibiotics. Acetyltransferases that inactivate aminoglycosides and chloramphenicol (Cantón, 2009), the gene cluster conferring glycopeptide resistance (Miao et al., 2012), and fosfomycin resistance determinants, found in clinically-relevant bacteria, can be traced down to respective producing organisms. The path of such genes (*i.e.*, HGT) is not difficult to imagine – although it is very hard to prove. However, many other resistance genes found in antibiotic-producing bacteria, are seldom found in pathogenic organisms. Aminoglycoside phosphotransferases, commonly found in producing-organisms and other soil bacteria, are not closely related to those found in clinical settings (Miao et al., 2012). Other resistance genes from producing bacteria have only recently been found in pathogens, and at very low rates (Cantón, 2009). From this limited evidence, it could be easy to assume that producing organisms have not played a significant role as a source of resistance determinants. However, it is important to realize that (a) many of those antibiotics are not only produced by the bacterial species they are industrially obtained from; and (b) there could be many more antibiotics in nature, with related chemical structure and/or mechanism of action, that have not been developed for clinical use or that have not been discovered. These are relevant considerations as, although some resistance genes are closely related to those in (known) antibiotic-producing bacteria, many more are not.

Being *Streptomyces* a genus capable of producing a wide variety of antibiotics, a collection of 480 strains of this genus was tested for antibiotic susceptibility. While the report (over)states the high prevalence of multi-resistance, it is important to notice that many of the phenotypes reported, such as trimethoprim, daptomycin and fosfomycin resistance, were present in all isolates, suggesting that such resistances are intrinsic to the genus. An arbitrary, 20-μg/mL concentration of all antibiotics was used for assessing "resistance", which is well above resistance breakpoints for most antibiotics upon gram-positives (and below the 32-μg/mL breakpoint for fosfomycin and chloramphenicol, and the 512 μg/mL for sulfonamides); without available breakpoint values for *Streptomyces*, interpretation of results as "resistance" is very difficult. The complete absence of aminoglycoside (streptomycin, neomycin and gentamicin) resistance was much more surprising; along with the inactivation of daptomycin, rifampicin, telithromycin and streptogramins, that have never been reported in clinical isolates (D'Costa et al., 2006). The paper introduced the notion of "resistome", but seems clear that much more work needs to be done to properly assess its magnitude.

2.2 RESISTANCE GENES OF CLINICAL RELEVANCE WITH KNOWN ENVIRONMENTAL ORIGINS

Other relevant resistance genes can be traced down to environmental bacteria, although not precisely to antibiotic-producers. The most cited example is the CTX-M extended-spectrum beta-lactamase (ESBL) family, which is now very common in clinical settings, and that seems to have originated from *Kluyvera* spp., a genus of soil bacteria (Cantón, 2009). This ESBL family emerged globally in clinical isolates in the 1990's, and is only distantly related to other ESBLs, such as the TEM and SHV families; however, it was found to be nearly identical to a chromosomal gene of *K. cryocrescens*, *K. georgiana* and *K. ascorbata*. The bla_{CTX-M} gene now common in pathogenic enterics is plasmid-borne, but is often accompanied by *Kluyvera* chromosomal, neighboring genes (Wright, 2012). *Kluyvera* is not known to produce antibiotics; what is an ESBL gene doing in this soil bacteria? How and when did this chromosomal gene gained access to the "genetic internet" created by HGT? Another example of a "resistance" gene that clearly originated from non-producing environmental bacteria, is *qnrA*, from *Shewanella algae*, an aquatic bacterial species. Qnr peptides protect bacterial topoisomerases from the action of fluoroquinolones, increasing the MIC but rarely up to a full-resistance phenotype (although their presence facilitate the emergence of mutations that lead to full-resistance). Quinolones are synthetic compounds, but other, naturally-occurring molecules are also topoisomerase-inhibitors (*e.g.*, novobiocin (Maxwell, 1993), microcins (Zamble et al., 2001)), which may explain the presence of these peptides in natural environments (Strahilevitz et al., 2009). Again, *qnr* genes in *Shewanella* are chromosomal, but are now found in plasmids of enteric bacteria. These two examples have several things in common: (a) bla_{CTX-M} and *qnrA* are genes only recently, but now commonly found in clinically-relevant bacteria, mostly in mobile genetic elements; (b) these genes originated from the chromosome of environmental, non-antibiotic-producing bacteria; (c) the role of these genes in antibiotic-free environments is not clear. They are also "poster boys" for the potential risks posed by the presence of antibiotic resistance determinants in environmental bacteria, which may be acting as reservoirs of resistance genes. However, important questions remain unanswered: Why these genes, but not the majority of those common in the soil (*e.g.*, those from D'Costa et al. paper, or the functional metagenomic experiment that found amino-glycoside and tetracycline resistance genes unrelated to previously reported sequences (Riesenfeld et al., 2004))? Is it only a matter of time? Are they already present in pathogens, but so far undetected? Or is there a functional barrier that prevents their mobilization or expression in clinically-relevant bacteria? Regarding this last question, the ecology *versus* phylogeny controversy around HGT (see section 1.2.1.4) comes to mind: would it be necessary for resistance genes in soil bacteria to simply share an ecological niche with pathogens in order to be adopted by the latter; or would one or several bacterial species be needed to bridge the phylogenetic gap between soil and pathogenic organisms? These questions need to be answered to adequately weight the relevance of the environmental "resistome". An interesting step in this direction was a functional metagenomic study of resistance genes in soil bacteria: there were nearly 3,000 genes conferring resistance when cloned into *E. coli* cells with, again, only less than 1% having sequence identity to resistance genes reported in human pathogens. Furthermore, "mobility elements" (*i.e.*, transposases or integrases) were significantly less and further away from such putative resistance genes, than they are in a sample of human pathogens (Forsberg et al., 2014). With the limitations of metagenomic studies enlisted in Chapter 1, and while neglecting many other mobility genes and mechanisms (only inter-molecular mobility genes were apparently addressed, and truly HGT mechanisms were not looked for), this study shows that resistance genes in soil bacteria do not seem to be prone to mobilize, at least not as they do among common pathogens. As will be analyzed in further chapters, this situation can change dramatically due to human-related selective pressures.

2.3 RESISTANCE GENES THAT WERE NOT "MEANT" FOR RESISTANCE

Leaving the realm of the chemical warfare scenario, and assuming a rather signaling role for antibiotics, it might be useful to recall what happens to other signaling molecules in nature. Neurotransmitters, for instance, are released by pre-synaptic neurons, accumulate in the synaptic cleft until reaching a threshold concentration that elicits the adequate response by the post-synaptic neuron . . . and are enzymatically inactivated (*e.g.*, by monoamine oxidases). This is crucial to ensure the timely activation and inactivation of neurotransmission. It could very well be the case for antibiotics acting as inter-cellular signaling molecules: inactivating enzymes both, from producing or receiving organisms, may exist to ensure the short-term life of the signal. This is, however, a mere speculation.

Some beta-lactamase enzymes, especially chromosomal ones (class C), seem to derive from enzymes involved in regulating peptidoglycan assembly, the PBPs; AmpC beta-lactamases are often inducible, and their regulatory system involve the recognition of peptidoglycan-autolysis products. The ability to produce beta-lactams and beta-lactamases may have evolved some 3.5 billion years ago in gram-negative bacteria (Medeiros, 1997). Class A beta-lactamases may have evolved some 2.4 billion years ago (Finley et al., 2013); TEM and SHV enzymes then diverged about 300–400 million years ago (Hall, 2007). It is therefore possible that present-day resistance mechanisms have evolved, perhaps very rapidly, from antibiotic biosynthetic pathways, again from known or unknown antibiotic producers. Such a situation could explain why, when taken out of their cellular context and expressed in laboratory strains (*i.e.*, functional metagenomics), many genes from environmental samples seem to confer "resistance", but have seldom been found in pathogens. The actual probability of such a jump remains to be elucidated.

Antibiotics both, natural and synthetic, are organic molecules that could be used as sources of carbon and/or nitrogen; many soil bacteria are capable of utilizing anti-biotics in this way, which would obviously involve both, the ability to grow in the presence of antibiotics, and the ability to degrade them. This has been termed "sub-sistence", instead of resistance (and genes involved in antibiotic catabolism have been termed the "antibiotic subsistome"; should this tendency persists, there would be a "nomenclaturome" grouping the many new arising names). Some organisms capable of utilizing beta-lactam antibiotics as sole source of carbon and nitrogen produce, among other enzymes, beta-lactamases. Additionally, pathogens such as *Salmonella* can subsist on one or several antibiotics, including synthetic sulfonamides. It is there-fore likely that some subsistence genes are at the origin of present-day resistance genes found in pathogenic bacteria (Dantas and Sommer, 2012). However, there is not much further evidence on this.

2.4 "RESISTANCE" GENES UNRELATED TO ANTIBIOTICS, AND NON-RESISTANCE GENES OF RELEVANCE

As reviewed in Chapter 1, a number of mechanisms conferring resistance to antibiotics are so unspecific that the selection pressures that led to their development are most likely unrelated to antibiotics; antibiotic resistance would then be a rather "accidental" side-effect. Unspecific efflux pumps, for instance, extrude a wide variety of xenobiotics, and their expression is also triggered by several non-antibiotic compounds. This con-trasts with the exquisite specificity of inactivating enzymes, drug-specific efflux pumps, or by-pass mechanisms, that protect against a very narrow group of compounds with similar chemistry or mechanism of action; when inducible, these mechanisms also respond exclusively to individual antibiotics (*e.g.*, chromosomal AmpC beta-lactamases, plasmid-encoded staphylococcal penicillinases, or *erm* genes). Most of these unspecific mechanisms can be classified as adaptive or even intrinsic resistance;

they often confer low-level, albeit multiple-drug resistance; and are unlikely to be mobilized horizontally, as are chromosomally-encoded. It would be surprising NOT to find these genes in environmental metagenomic screenings (although it is a bit surprising to find them in resistance gene databases); all these genes point to is to the existence of the known bacterial bearers in the studied samples. In the absence of an additional, human-originated stress, these mechanisms are likely to keep playing their natural role; but sudden environmental changes can also turn them into potential threats: (a) transient expression of these mechanisms caused by a new inducing stimulus can protect the carrier and co-select for any other dangerous trait, modifying the equilibrium of the bacterial population; (b) constitutive mutants, normally at a disadvantage, could become prevalent should a selective pressure, such as an antibiotic or disinfectant appears; (c) given the right conditions (*i.e.*, gene recruitment into a mobile element, mobilization/stabilization/expression into a pathogenic or commensal bacteria, acquisition of such a new host by an animal or human liable to be treated with antibiotics) such mechanisms could potentially result in antibiotic treatment failure. There is, however, no evidence that this has ever happened.

Resistance to other toxic agents are also important to understand the spread of antibiotic resistance genes. Heavy metals, for instance, are often toxic for bacteria at concentrations related to human activities. Resistance to these ions is not rare in bacteria, and such determinants are often found along with antibiotic resistance genes (and, furthermore, within mobile genetic elements). The most extensively studied is mercury resistance: *mer* genes are organized in an operon under the regulation of MerR, and encode an Hg^{2+} transport protein, a reductase and, some times, an organomercuric lyase. But there is also plasmid-mediated resistance to arsenic, antimony, cadmium, silver, chrome and copper, among others. The selection of these resistances by industrial pollution and the co-selection enabling of antibiotic resistance genes have been discussed for more than 30 years (Misra et al., 1984). The human-related release of these metals into the environment will be discussed in the next chapter.

There are genes other than resistance ones (to antibiotics or other xenobiotics) that are particularly relevant for the purposes of this book: mobility genes. Genes enabling the movement of gene cassettes and transposons between DNA molecules, and those enabling HGT, have been of crucial importance for the spread of resistance determinants. It is obvious that they existed before and independently of the presence of antibiotics; main HGT mechanisms were actually discovered *before* the "antibiotic era", and conjugative plasmids were common in strains of the Murray Collection (Hughes and Datta, 1983), a set of clinical isolates predating the introduction of antibiotic into clinical use. A wide variety of integrons were detected in samples from different environments, including some mostly undisturbed and without antibiotic exposure (Nield et al., 2001), indicating that this form of mobilization was also common before the human use of antibiotics; environmental Betaproteobacteria might be the original bearers of class 1 integrons, today very commonly found in pathogens carrying antibiotic resistance cassettes (Gillings et al., 2008). Insertion sequences that give rise to composite transposons are common in non-pathogenic bacteria, rearranging genotypes unrelated to antibiotic resistance, such as those involved in the catabolic degradation of organic pollutants (Di Gioia et al., 1998).

It is clear that all ingredients were already present within bacterial cells much before the discovery of antibiotics: resistance and proto-resistance genes, and mobility

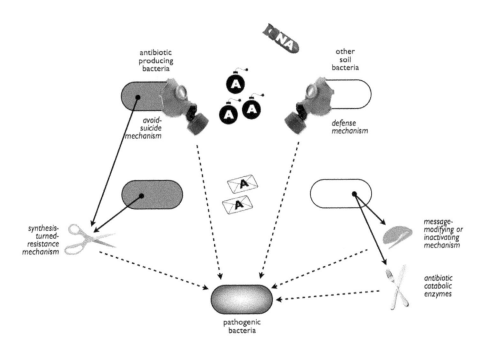

Figure 2.1 **Naturally-occurring antibiotics and resistance genes: a graphic summary.** In a chemical warfare scenario, antibiotics (A, as hand bombs) are produced by some bacteria in order to attack their neighboring fellow soil bacteria; these antibiotic-producers need defense mechanisms to "avoid suicide" (the gas mask). There could be many unknown bacteria producing known antibiotics (and respective unknown defense mechanisms). Although all efforts to detect naturally-produced antibiotics directly in soil samples have failed, it is conceivable that, in some scenarios, these chemical weapons are concentrated enough to select for defense mechanisms (again, the gas mask) in other soil bacteria. These defense mechanisms could protect against known and unknown-but-related antibiotics in the soil; they could also protect against non-antibiotic agents present in the environment (NA, as dropping bomb). Conceiving antibiotics rather as chemical messages (A, in a letter envelope), the producing organism would have a complex synthesis pathway, including modifying enzymes that could turn into resistance mechanisms (this could also be the case in the warfare scenario). Other soil bacteria could modify or inactivate these messages (a letter opener), or even use them as food with appropriate catabolic enzymes (a fork and knife). There are known examples within each group of these genes that have been detected in pathogenic bacteria, acting as resistance determinants; the pathways of mediating HGT events are not known.

genes both, for inter-molecular and inter-cellular transfer. Different selective pressures were also at play before the "antibiotic era": toxic synthetic compounds and metal ions were released into the environment early in the industrial age; some, such as mercury, were even used as disinfectants. When massive amounts of antibiotics and other xenobiotics started to reach the environment, the modification, selection and mobilization – not necessarily in that order, of resistance genes into human pathogens was swift and caused said "antibiotic era" to be so short-lived.

There are many reports of resistance genes found in environmental bacteria, most often obtained from metagenomic approaches; these will be analyzed in Chapter 4.

This chapter only addressed the non-human-related presence of antibiotics in the environment, and the few known examples of resistance genes present in pathogenic bacteria that can be traced down to antibiotic-producers or other ancient, perhaps neighboring bacteria that were under the selective pressure of naturally-occurring antibiotics. All other reports, of resistant bacteria and resistance genes found in all sorts of environments, fall into different categories, either as human-related contaminants, or as natural potential sources of health risks.

REFERENCES

Baquero, F., Negri, M. C., Morosini, M. I. & Blázquez, J. (1998) Antibiotic-selective environments. *Clin. Infect. Dis.*, 27 (suppl. 1), S5–S11.

Cantón, R. (2009) Antibiotic resistance genes from the environment: a perspective through newly identified antibiotic resistance mechanisms in the clinical setting. *Clin. Microbiol. Infect.*, 15 (suppl. 1), 20–25.

D'Costa, V. M., McGrann, K. M., Hughes, D. W. & Wright, G. D. (2006) Sampling the antibiotic resistome. *Science*, 311, 374–377.

Dantas, G. & Sommer, M. O. A. (2012) Ecological and clinical consequences of antibiotic subsistence by environmental microbes. In Keen, P. L. & Montforts, M. H. M. M. (Eds.) *Antimicrobial resistance in the environment*. Hoboken, John Wiley & Sons.

Di Gioia, D., Peel, M., Fava, F. & Wyndham, R. C. (1998) Structures of homologous composite transposons carrying *cbaABC* genes from Europe and North America. *Appl. Environ. Microbiol.*, 64, 1940–1946.

Finley, R. L., Collignon, P., Larsson, D. G. J., McEwen, S. A., Li, X. Z., Gaze, W. H., Reid-Smith, R., Timinouni, M., Graham, D. W. & Topp, E. (2013) The scourge of antibiotic resistance: the important role of the environment. *Clin. Infect. Dis.*, 57, 704–710.

Forsberg, K. J., Patel, S., Gibson, M. K., Lauber, C. L., Knight, R., Fierer, N. & Dantas, G. (2014) Bacterial phylogeny structures soil resistomes across habitats. *Nature*, 509, 612–616.

Gillings, M., Boucher, Y., Labbate, M., Holmes, A., Krishnan, S., Holley, M. & Stokes, H. W. (2008) The evolution of class 1 integrons and the rise of antibiotic resistance. *J. Bacteriol.*, 190, 5095–5100.

Hall, B. G. (2007) Predicting the evolution and emergence of new antibiotic resistance genes: an important element in developing antibiotics and antibiotic therapeutic policy. In Amábile-Cuevas, C. F. (Ed.) *Antimicrobial resistance in bacteria*. Wymondham, Horizon Bioscience.

Hochberg, M. E. & Jansen, G. (2015) Assessing resistance to new antibiotics. *Nature*, 519, 158.

Hughes, V. M. & Datta, N. (1983) Conjugative plasmids in bacteria of the "pre-antibiotic" era. *Nature*, 302, 725–726.

Laskaris, P., Gaze, W. H. & Wellington, E. M. H. (2012) Environmental reservoirs of resistance genes in antibiotic-producing bacteria and their possible impact on the evolution of antibiotic resistance. In Keen, P. L. & Montforts, M. H. M. M. (Eds.) *Antimicrobial resistance in the environment*. Hoboken, John Wiley & Sons.

Ling, L. L., Schneider, T., Peoples, A. J., Spoering, A. L., Engels, I., Conlon, B. P., Mueller, A., Schäberle, T. F., Hughes, D. E., Epstein, S., Jones, M., Lazarides, L., Steadman, V. A., Cohen, D. R., Felix, C. R., Fetterman, K. A., Millett, W. P., Nitti, A. G., Zullo, A. M., Chen, C. & Lewis, K. (2015) A new antibiotic kills pathogens without detectable resistance. *Nature*.

Maxwell, A. (1993) The interaction between coumarin drugs and DNA gyrase. *Mol. Microbiol.*, 9, 681–686.

Medeiros, A. A. (1997) Evolution and dissemination of b-lactamases accelerated by generations of β-lactam antibiotics. *Clin. Infect. Dis.*, 24 (suppl. 1), S19–S45.

Miao, V., Davies, D. & Davies, J. (2012) Path to resistance. In Keen, P. L. & Montforts, M. H. M. M. (Eds.) *Antimicrobial resistance in the environment.* Hoboken, John Willey & Sons.

Misra, T. K., Silver, S., Mobley, H. L. T. & Rosen, B. P. (1984) Molecular genetics and bio-chemistry of heavy metal resistance in bacteria. In Tashjian, A. H. (Ed.) *Molecular and cellular approaches to understanding mechanisms of toxicity.* Boston, Harvard School of Public Health.

Nield, B. S., Holmes, A. J., Gillings, M. R., Recchia, G. D., Mabbutt, B. C., Nevalainen, K. M. H. & Stokes, H. W. (2001) Recovery of new integron classes from environmental DNA. *FEMS Microbiol. Lett.,* 195, 59–65.

Riesenfeld, C. S., Goodman, R. M. & Handelsman, J. (2004) Uncultured soil bacteria are a reservoir of new antibiotic resistance genes. *Environ. Microbiol.,* 6, 981–989.

Strahilevitz, J., Jacoby, G. A., Hooper, D. C. & Robicsek, A. (2009) Plasmid-mediated quinolone resistance: a multifaceted threat. *Clin. Microbiol. Rev.,* 22, 664–689.

Waksman, S. A. & Woodruff, H. B. (1940) The soil as a source of microorganisms antagonistic to disease-producing bacteria. *J. Bacteriol.,* 40, 581–600.

Wright, G. D. (2012) Antibiotic resistome: a framework linking the clinic and the environment. In Keen, P. L. & Montforts, M. H. M. M. (Eds.) *Antimicrobial resistance in the environment.* Hoboken, John Wiley & Sons.

Zamble, D. B., Miller, D. A., Heddle, J. G., Maxwell, A., Walsh, C. T. & Hollfelder, F. (2001) *In vitro* characterization of DNA gyrase inhibition by microcin B17 analogs with altered bisheterocyclic sites. *Proc. Natl. Acad. Sci. USA,* 98, 7712–7717.

Chapter 3

Human-related release of antibiotics into the environment

Contrasting with the minute quantities of antibiotics released by producing organisms, huge amounts of industrially-produced antibiotics have been released into the environment since the beginning of the "antibiotic era". This can go from the drugs excreted by medicated humans and other animals in their urine and feces, to the application of antibiotics upon trees and grasses, to the dumping of industrial by-products into water bodies. These and other variants will be discussed in this chapter. However, there is a long list of largely unknown or unquantified instances of environmental release of antibiotics, of which there is no available data, neither on their magnitude nor on their impact. Many of these have to do with the poor sanitary conditions and urban infrastructure, lack of regulatory and enforcement frameworks, and corruption, that are typical of non-developed countries. In fact, corruption and poor governance correlate much better to increased bacterial resistance amongst pathogens, than clinical antibiotic usage, at least in European countries (Collignon et al., 2015). Among the things that do happen, but that may or may not have an impact on the emergence and spread of antibiotic resistance are, in the urban setting:

- The excretion of antibiotics by medicated humans and pets directly into the streets, by open urination and defecation.
- The running of urban sewage – carrying antibiotics from houses and hospitals, in open, exposed channels, and/or directly into the streets, due to sewage leaks or flooding.
- The dumping of antibiotics and antibiotic-containing garbage, from households, clandestine drugstores, and hospitals, into common municipal waste, carried in open, exposed trucks, and delivered to open, exposed fields; leachates from these fields can potentially carry antibiotics deep into the underlying ground.
- The exposure of urban fauna (*e.g.*, insects, mice, rats, birds, stray dogs and cats) to all these antibiotic sources, and the consequent exposure of their microbiota.

And in the rural setting:

- The administration of antibiotics to animals, mainly for therapy or "prophylaxis" rather than "growth promotion", in small-scale farming activities.
- The usage of antibiotic-containing wastewater and manure into small-scale, self-contained agricultural activities.
- The wider distribution of meat and vegetables from the sources above, due to the growing fad of "non-industrial" foodstuff.

As mentioned before, the actual amount of antibiotics released by these practices is unknown, as they are unregulated, often clandestine situations. But is seems very likely that, for instance, rats, stray dogs, or even people foraging a dumpster, may encounter hospital garbage contaminated with antibiotics; and that resistant bacteria may be selected from the microbiota of these people or animals.

As to well-recognized sources of antibiotics released into the environment both, in developed and non-developed countries, there are many documented examples. It is not within the purview of this book to make an exhaustive review of such cases, but only to highlight the main types of activities resulting in antibiotic release, and to provide some examples as to their magnitude (Figure 3.1). Some of these sources involve antibiotics that are excreted by humans or other animals receiving such drugs for therapeutic or other purposes. It is therefore important to keep in mind that excretion differs between antibiotics (and probably between animal species; the following paragraph would only include information about excretion in humans), so that the actual amount of antibiotic released to the environment along with urine or feces, would not only depend on the amount taken, but also on the excretion route. Most antibiotics are mostly excreted in active form; important exceptions are chloramphenicol, metronidazole and, to some extent, linezolid. Most antibiotics are mostly (>75%) excreted in the urine: beta-lactams (except for ceftriaxone), aminoglycosides, quinolones, nitrofurantoin, sulfonamides and glycopeptides. On the other hand, macrolides, tetracyclines and fusidic acid, for instance, are excreted mainly in the feces; and fosfomycin and rifampicin (and ceftriaxone) are excreted somewhat 50/50. There are two further considerations: (a) these excretion routes refer only to the absorbed antibiotic, when administered orally; however, absorption rarely amounts to 100% of the ingested dose: 13% of nitrofurantoin is not absorbed orally, nor is 35% of ciprofloxacin, 45% of clarithromycin, 65% of erythromycin or 72% of fosfomycin; and (b) the non-absorbed fraction, along with that returning to the intestinal tract (the biliary-fecal route), and the small fecal volume, usually make for fully inhibitory concentrations in the feces of treated animals and humans.

Although the main discussed consequence of the release of antibiotics in the environment is the influence these compounds exert upon bacterial populations, especially the likelihood of them selecting for resistance, there are reports of other kinds of organisms affected by the presence of small concentrations of antibiotics (Segura et al., 2009). A recent report shows that tetracycline, one of the most persistent antibiotics found in waters and soils, can affect many eukaryotic organisms, even delaying plant growth (Moullan et al., 2015). These effects would not be further analyzed in this text.

3.1 AGRICULTURAL USE OF ANTIBIOTICS

The non-clinical use of antibiotics is the most contentious issue regarding bacterial resistance. Practices grouped under this label have been blamed for the dramatic rise in resistance among clinically-relevant bacteria; unfortunately, as with the link between air pollution and climate change – and some years before, between smoking and lung cancer, the link between agricultural use of antibiotics and rising resistance, is difficult to establish. This, along with the formidable economic interests behind the practices, that fuel the lobby for political inaction, and misinformation generation, result in

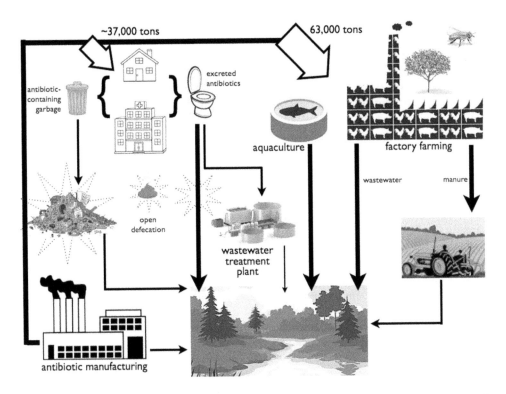

Figure 3.1 **Antibiotic pathways into the environment.** Pharmaceutical companies (bottom left) produce antibiotics used on food animals (63,000 tons per year) and on humans and pets (~37,000 tons per year, calculating standard doses at 500 mg), that are then released directly, or after being metabolized and excreted, into the environment; some are also dumped along with industrial wastewater. In urban settings in developed countries, excreted antibiotics end up mostly in wastewater which in turn goes into treatment facilities before being released into rivers or other water bodies. In non-developed cities, wastewater is rarely treated before release, and antibiotics also end up in municipal garbage in open landfills, from which they can reach foraging humans and animals, and be transported to soils and waters through leaching and runoff (dotted-line star); open defecation and urination, and open-channel sewage, also expose unknown amounts of antibiotics to the urban environment. Although wastewater treatment reduces antibiotic concentrations (while, perhaps, converting some antibiotics into more potent derivatives), detectable amounts of many antibiotics are still released to water bodies. In rural settings, antibiotics are used in factory farming, aquaculture, vegetables and bees. Factory farming wastewater is often dumped into water bodies, and antibiotic-containing manure is massively used as fertilizer; antibiotics may remain and accumulate in soils, or can be transported to water bodies as leachates or runoff. Antibiotics used in aquaculture are usually applied directly at natural water bodies, if using nets or cages; or can leach to the soil and then to groundwater, if using artificial ponds. Although antibiotics are found in sub-inhibitory concentrations in most of these instances, they do have effects upon environmental microbial communities, with consequences that are hard to foresee.

haphazard measures that are at best incomplete. A clear example of this is the FDA Guidance for Industry #213 (analyzed by the PEW Charitable Trusts (2014)), intended to reduce antibiotic abuse in livestock, issued in December, 2013: veterinarians are now to oversee the addition of antibiotics to feed and water (which may have an impact similar to that of having psychologists overseeing torture practices by the CIA); and drug manufacturing companies are to remove labeling of "weight gain" indications, although drugs intended for "disease prevention" still include the "maintaining weight gain" as an expected result of their use, and dosage intended for each purpose frequently overlap.

The use of antibiotics as "growth promoters" on farm animals started accidentally, during the late 1940's, with the use of byproducts of tetracycline production as additive for chicken food, intended as a vitamin B source. As animals grew faster, among other productivity advantages, the practice of supplementing animal food with subtherapeutic doses of antibiotics rapidly extended. When multi-resistant bacteria causing food-borne infections became a problem, a committee was formed in the UK to analyze the available evidence on the risks of this use of antibiotics; this committee recommended, in 1969, a series of restrictions upon the use of antibiotics as animal food additives. Although even those restrictions were too lax, feeble implementation and recurrent attack from the pharmaceutical and farming industries resulted in their rapid dilution during the 1970's and 1980's (Edqvist and Pedersen, 2001).

In 2010, more than 63,000 tons of antibiotics were used worldwide for food animal production, half of them consumed in China (23%), US (13%), Brazil (9%), India (3%) and Germany (3%); and an estimate of 105,000 tons by 2030, a third of the increase due to shifts to intensive farming systems (there is also data indicating that the "growth promotion" activity of antibiotics is declining: in the 1950's, 5–10 ppm of tetracycline were used, while now 50–200 ppm are needed (Levy, 2002)). Average yearly consumption is estimated at 172 mg per kilogram of produced pig, 148 mg/kg of chicken and 45 mg/kg of cattle (Van Boeckel et al., 2015). Although some figures of antibiotic usage in animals vary widely between 50% and 80% of all produced antibiotics, actual data is scarce, as regulation of this usage is very variable. In the US, in 2012, 10,000 tons of antibiotics were sold for use in animals, while only 3,290 tons were sold for human use. Nearly 30 different antimicrobial classes are used in animals, most of them also used in humans; the top three classes used in animals, by sales, were macrolides, penicillins and tetracyclines, each in the order of USD 500 million annually (Laxminarayan et al., 2015). It is important to consider that for each human receiving antibiotics per year, there are 30 animals also receiving antibiotics; and some individual animals can excrete up to 100 times more feces than a single person (Levy, 2002).

Antibiotics used on farm animals end up in the environment through many different ways. If at large, industrial facilities in developed countries, liquid waste (slurry) is likely disposed along with wastewater, which then undergo a series of treatments (see below), with only small amounts leaking into the environment, as runoff and leachates. But in smaller scale production facilities and, especially, in non-developed countries, liquid waste is dumped into water bodies without any previous treatment. For instance, the Jiyun river, running through a livestock production area in China, averages concentrations (in ng/L) up to 16 for tetracyclines (with peaks of 100), 28 for fluoroquinolones (peaks of 102), 148 for sulfonamides (peaks of 450) (Zhang et al., 2014). As to the solid waste (i.e., manure), including antibiotic-containing feces, vegetal material used as bedding, impregnated with feces and urine, and remnants of antibiotic-containing food, it is very often used as fertilizer. About 132 million tons of manure are produced yearly in the US alone, with a typical antibiotic concentration of 1–10 mg/kg or L (and up to 200 mg/kg or L); manure is applied to 9.2 million ha (the approximate surface of Portugal) in the US (Dolliver, 2007). In pig manure from Germany, antibiotic concentrations can reach, in mg/kg, 50.8 for chlortetracycline, 46 for tetracycline, and 38 for sulfamethazine; although median concentrations are only in the tenths of mg/kg for these and other tetracyclines and sulfonamides. Degradation of antibiotics in manure is dependent on temperature; at 20–22°C, sulfamethoxazole, enrofloxacin, florfenicol and trimethoprim are degraded up to 70–80% in 16 weeks; but most other sulfonamides suffer little or no degradation under these conditions (Harms and Bauer, 2012).

In addition to the use as "growth promoters" and prophylaxis upon typical farm animals, antibiotics are also used for prophylaxis and treatment in beekeeping, aquaculture, and even upon fruit trees and golf course grasses. While some of these practices only use a tiny fraction of the antibiotics consumed by typical farm animals, these antibiotics are directly dumped into soils or water bodies, hence reaching higher concentrations and potentially exerting a stronger selective pressure. Furthermore, much less effort have been devoted to analyze the impact of these instances:

–   Aquaculture is a rapidly-growing trend for food production worldwide; by 2004, it produced 59.4 million tons, with a 7.1% average annual growth rate; 91.5% of this production comes from the Asia-Pacific region (69.9% from China alone). Mollusks and aquatic plants account for 51% of global aquaculture, finfish for most of the 43% of freshwater culture, and shrimps for two-thirds of the 6% brackishwater production. Culture systems are diverse, from freshwater earthen ponds, to closed or semi-closed recirculating systems, to flow-through tanks, net pens and cages. Antibiotics are often used against infections by *Aerococcus*, *Aeromonas* or *Edwardsiella* spp. Methods for antibiotic treatment depend on the culture system: if in tanks, the water volume is reduced and the antibiotic added for a 1–2 h exposure; tanks can also be immersed into a solution for a few seconds or minutes (dip) or a few hours (short bath); a "flush" involves a concentrated solution that enters, flows quickly and goes through the effluent pipe. Antibiotics can also be added to the feed. The amount of antibiotics used in this way varies widely, from 2 g per ton (Sweden) to 157 g/ton (Canada) to 700 g/ton (Vietnam); a simple multiplication would put the worldwide use of antibiotics for aquaculture somewhere between 120 and 42,000 tons yearly; in the US in the mid-1990's, 92–196 tons of antibiotics were used in aquaculture (Laxminarayan et al., 2015). Antibiotics most commonly used in aquaculture are oxytetracycline, co-trimoxazole (or sulfadimethoxine/ormetoprim), quinolones (oxolinic acid and flumequine) and florfenicol; however, aminopenicillins, erythromycin, streptomycin, neomycin and nitrofurans are also used in this way (FAO et al., 2006). A particularly unfriendly report (as it uses ppt concentration units, hardly compatible with any other report) puts sulfonamides, trimethoprim and fluorquinolones in the 10^6-ppt concentration range in Indochina shrimp ponds (Suzuki and Hoa, 2012); if my math is not wrong, that would be about μg/L ranges.

- Oxytetracycline has been used by beekeepers since the 1950's to control larval foulbrood diseases caused by *Melissococcus pluton* and *Paenibacillus larvae* (Tian et al., 2012). It is applied both, by dusting colonies, and in a sugar syrup used for feeding, at doses of 250 mg per colony every 3–4 days.
- Streptomycin is used mostly to control fire blight (an infection of apples and pears, caused by *Erwinia amylovora*) in the US, Israel, New Zealand, Canada and Mexico and, occasionally, in Germany, Austria and Switzerland; oxytetracycline is used for the same purposes and against bacterial spot (infections by *Xanthomonas arboricola* upon peaches and nectarines) in the US, Mexico and Central America; gentamicin is used in Mexico and Central America against infections by *E. amylovora*, *Pectobacterium*, *Pseudomonas*, *Ralstonia* and *Xanthomonas*; oxolinic acid, a quinolone, is used against streptomycin-resistant *E. amylovora*, and in Japan against panicle blight of rice, caused by *Burkholderia glumae*. In the US alone, more than 16,000 kg of oxytetracycline and streptomycin are sprayed on apples, peaches and pears per year; while this only amounts to 0.12% of the total antibiotics used in animal agriculture (Stockwell and Duffy, 2012), these are directly applied to plants and soils, instead of being, at least temporarily, confined to the intestinal tract of animals. A further use of antibiotics upon plants can be the best example of reckless abuse: oxytetracycline solution (50 gallons per 1000 square feet) is applied to treat turfgrasses affected by bacterial wilt, caused by *Xanthomonas campestris* pv. *graminis*.

3.2 WASTEWATER

Antibiotic usage in humans varies widely from country to country, and there is no available information for some parts of the world, such as Africa and many Asian countries. According to the most recent and complete report available, there are countries like China and Brazil where antibiotic consumption was 1–8 standard units (pill, capsule or ampoule) per person in 2010, contrasting with 75–625 in Australia and New Zealand; the US and many European countries fall within the 20–28 units/person category. A total of 73.6 billion antibiotic units were used in 2010 in 71 countries were sales data were available, being the most commonly used antibiotic groups, broad-spectrum penicillins (about 24 billion units), cephalosporins (~18 billion units), macrolides (~8 billion units), fluoroquinolones (~7 billion units), and trimethoprim and tetracyclines (~5 billion units each) (Van Boeckel et al., 2014). The dose contained in each "standard unit" varies from drug to drug but, just to have an idea of the amount of antibiotics actually used for clinical purposes, if an average of 500 mg per unit is a valid approximate, then about 37,000 tons of antibiotics are consumed by humans. As mentioned before, most antibiotics are excreted in active form in the urine and/or feces; and with about half of the human population now living in cities, perhaps something around 15,000 tons of antibiotics are dumped yearly into the sewage. That is 40,000 kg per day. Wastewater carries A LOT of antibiotics. While these figures are gross generalizations, a peculiar estimate of 305 tons of excreted antibiotics released into German wastewater in 1998 (68% of them being beta-lactams and 25% sulfonamides (Kümmerer and Henninger, 2003)) indicates that crude calculations presented above are not entirely clueless.

3.2.1   Early release of antibiotics to the sewage

Antibiotics are rapidly diluted in wastewater, keeping inhibitory concentrations for brief periods of time, and acting upon bacteria in the sewage. While hospitals are concentrated hotspots of antibiotic usage, the bulk of antibiotics are used by outpatients: 75–95% of antibiotics used clinically in the US, Germany and UK are consumed in the community (Kümmerer, 2003). Therefore, most of excreted antibiotics are flushed in house or office toilets in urban settings, and become diluted in the 150–600 L of water used daily per person that also end up in the sewage. Nevertheless, hospital effluents contribute with highly concentrated antibiotics to wastewaters: 0.7–124 µg/L ciprofloxacin, and 20–80 µg/L ampicillin have been detected in hospital effluents (Kümmerer, 2003). German estimations of antibiotic concentrations at hospital effluents, based on consumption, excretion, and water usage rates, were in 1998 (in µg/L): 355–476 for penicillins, 300–494 for cephalosporins, 1.5–13.5 for aminoglycosides, 0.6–1.1 for tetracyclines, 2.1–3.8 for macrolides, 0.1–20.7 for glycopeptides, 16.9–38.0 for quinolones, 8.7–24.9 for carbapenems, and 18.2–92.8 for sulfonamides (Kümmerer and Henninger, 2003). In sewage drains near hospitals at Lahore City, Pakistan, antibiotics were found at similar concentrations (in µg/L): 4.6 for sulfamethoxazole, 3.2 for oxytetracycline, 2.2 for trimethoprim, 1.1 for lincomycin, and 1.0 for quinolones (Khan et al., 2013).

Antibiotics known to act on "concentration-dependent" basis (*i.e.*, a brief exposure to a high concentration can kill susceptible bacteria rapidly, or inhibit them for a long time), such as fluoroquinolones and aminoglycosides, would exert a larger selective pressure in the earlier moments after their release, than those that are "time-dependent", such as beta-lactams or macrolides, that need to keep concentrations above the MIC for most of the time to have a sustained effect upon bacterial viability. Once in wastewaters, antibiotics have different fates, depending on their physicochemical properties, as well as those of the diluting media *per se* (*i.e.*, pH, temperature, other solutes, light exposure). Some antibiotics, such as aminoglycosides and quinolones, are particularly stable in solution; while others, such as beta-lactams, tend to hydrolyse quickly. Being constantly released into wastewater, antibiotics that are not degraded rapidly can be considered persistent contaminants.

3.2.2   Along the way to treatment or final release

An extensive review of antibiotic concentration in wastewaters report median values in the two- to three-digit ng/L: 60 for lincosamides, 110 for macrolides, 205 for quinolones, 270 for trimethoprim, 300 for beta-lactams, 330 for sulfonamides, and 530 for tetracyclines. However, peaks are in the µg/L and even mg/L (or µg/mL, that is, around MICs) range: 920 mg/L for tetracyclines, 31 mg/L for quinolones, 1 mg/L for sulfonamides. Detection of metabolites or degradation products is also important, as many of those retain resistance-induction capabilities; median concentrations of 7.6 mg/L beta-lactam by-products, 1.4 mg/L tetracycline by-products, and 450 ng/L macrolide by-products have been reported in wastewaters (Segura et al., 2009).

Industrial wastewater from antibiotic manufacturing facilities can contain significant amounts of antibiotics. In developed countries such facilities often have their own treatment plants, so the effluent released to the environment may have low antibiotic concentrations; but in non-developed countries, industrial wastewater is released directly to water bodies. Antibiotics found at high concentrations in the vicinity of pharmaceutical facilities in Pakistan were (in μg/L): 49 for sulfamethoxazole, 28 for trimethoprim, 27 for oxytetracycline, 8 for levofloxacin, 7.3 for ofloxacin, and 6.2 for ciprofloxacin (Kümmerer, 2003). Up to 11.4 μg/L of ampicillin, and 155 ng/L of tetracycline were detected in the Almendares River, Cuba, receiving among other things, untreated waste from pharmaceutical factories (Graham et al., 2011). Nearly 1 mg/g of ciprofloxacin was found in organic matter of an industrial treatment plant near Hyderabad, India (Kümmerer, 2003).

3.2.3 Wastewater treatment plants

Developed countries collect urban sewage into treatment plants, before water is released into rivers, ponds or oceans. The role these treatment plants play in the dispersal of antibiotics and antibiotic resistant bacteria will be further discussed. However, it is important to realize that most cities, those in non-developed countries, do not have such facilities, hence wastewater is released without further treatment. For instance, the major treatment plant in Mexico City, one of the most populated cities in the world, receives only about 8% of the generated wastewater, the rest being discharged raw into the environment. Approximately 80–90% of wastewater in non-developed countries is not treated before their release into water bodies (Corcoran et al., 2010); more than 83% of the world population lives in non-developed countries now, and nearly 87% will by 2050. Therefore, most of the wastewater generated worldwide is released without treatment into the environment.

Antibiotics concentrations at wastewater treatment plants likely vary depending on the volume received, the population served, per-capita water consumption, and the antibiotic consumption rates (which are in turn affected by seasonal variation, outbreaks, etc.). Most detectable antibiotics reach wastewater treatment plants at concentrations in the ng/L range; treatment procedures tend to reduce concentrations: tetracyclines are removed by adsorption, beta-lactams by hydrolysis; but erythromycin and ciprofloxacin are recalcitrant (Pruden et al., 2013). Some procedures can actually increase antibiotic concentration at the outlet: the Oxford treatment plant, using activated sludge as the main biological treatment step, increased concentration of cefotaxime, from below detection limits to 51 ng/L; doxycycline from 60 to 121 ng/L; and trimethoprim from 70 to 73 ng/L. At the Oxford and Benson treatment plants, detectable antibiotic concentrations ranged, at the inlet, from 18 ng/L (cefotaxime) to 2320 ng/L (ofloxacin); and at the outlet, from 14 ng/L (ciprofloxacin) to 244 ng/L (erythromycin). All these values were obtained on November, 2009, at the peak of the influenza pandemic, during which antibiotic consumption also peaked (Singer et al., 2014).

While it is often considered that wastewater treatment reduces the concentration of antibiotics, an interesting side-effect may complicate the issue: some of the chemical reactions that antibiotics undergo during wastewater treatment, render new antibacterially active compounds that may go undetected by routine analysis, but still exert selective pressure (Keen and Linden, 2013). For instance, the chlorination step at many wastewater treatment plants can have the unexpected result of converting doxycycline into a new compound with increased antibacterial capabilities. Therefore, the actual antimicrobial load of wastewaters could be larger than usually estimated.

3.2.4 Release from treatment plants

Downstream of wastewater treatment plants, detectable antibiotic concentrations are in the single- to double-digit ng/L; during the peak of the 2009 influenza pandemic, rivers downstream of UK treatment plants had average concentrations (in ng/L) of 53 for erythromycin (with a 448 peak), 37 for clarithromycin (with a 292 peak), 28 for cefotaxime, 21 for azithromycin, 20 for ciprofloxacin, 17 for sulfamethoxazole, 16 for oxytetracycline, 11 for ofloxacin, 8 for norfloxacin, 8 for trimethoprim, and 6 for doxycycline. Two years later, azithromycin, cefotaxime, norfloxacin and ofloxacin were below detectable range (Singer et al., 2014). In the UK and other European countries, effluent concentrations were (in ng/L) of 80–832 for erythromycin, 70–438 for sulfonamides, 128–271 for trimethoprim, 20–95 for ciprofloxacin (Johnson et al., 2015).

The presence of antibiotics in natural waters range widely, from very high concentrations downstream of wastewater treatment plants and drug manufacturing facilities, to undetectable concentrations in "pristine" environments. A review reported median concentrations ranging from 11 ng/L of beta-lactams and macrolides, to 12 ng/L of sulfonamides (with maximum reported values of 1.6 mg/L), and 192 ng/L of tetracyclines (with maximum of 0.7 mg/L). Again, degradation products can be found in concentrations as high as 4 mg/L of beta-lactams and 11 µg/L of tetracyclines (Segura et al., 2009). Another review places antibiotic concentrations also in two- to three-digit ng/L, from 20 for trimethoprim, to 690 for chlortetracycline; important exceptions are erythromycin, with 1,700; sulfadiazine, with 4,130; lincomycin, with 21,100; and oxytetracycline, with 32,000 (Kemper, 2008). It is important to notice that none of these concentrations is near MICs of clinically-relevant bacteria; this is not to say that they do not exert effects upon microbial communities.

Fortunately, antibiotics do not seem to be present in drinking water. Among the very few reports finding them, a review includes finding 2–5 ng/L of macrolides and quinolones, and 0.4–0.5 ng/L of sulfonamides (Segura et al., 2009); and an informal report of 8 ng/L of amoxicillin (and 19 ng/L of derivative 6-aminopenicillanic acid) in tap water from Nanjing, China (Huang et al., 2015).

As can be seen, antibiotics are, with notable exceptions, rapidly degraded and/or diluted so that water and soil concentrations are several orders of magnitude below MICs, even in the vicinity of urban or otherwise antibiotic-using settings. Many people believed, and still argue, that these sub-inhibitory concentrations pose therefore no risk for the selection or development of fully-resistant bacteria; this is definitely not true, as will be analyzed in Chapter 5.

3.3 SOILS

Human-made antibiotics can reach the soil in a number of ways. Most of the antibiotic burden is caused by the application of manure, which can amount up to kilograms per hectare; but grazing livestock receiving antibiotics also release them in their urine

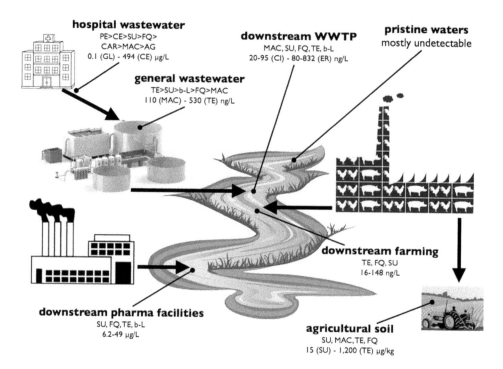

Figure 3.2 **Antibiotic concentrations in the environment.** Most abundant sources, most commonly found antibiotics (AG, aminoglycosides; b-L, beta-lactams, CAR, carbapenems; CE, cephalosporins; Cl, ciprofloxacin; ER, erythromycin; FQ, fluoroquinolones; GL, glycopeptides; MAC, macrolides; SU, sulfonamides; TE, tetracyclines), and selected examples of detected concentrations in water and soil.

and feces to the soil. Water-carried antibiotics, either from streams or as leachates or runoff from farming facilities, may end up adsorbed in soils; the rate at which they are adsorbed depend on the physical-chemical characteristics of each drug. Antibiotics can then be fixated (and accumulate, if added at a rate higher than they are degraded), or can mobilize to ground waters by leaching, or to surface waters by erosion. The fate of each antibiotic depends on its own physical-chemical properties, and those of the soil itself, so it is very variable; degradation products, which also vary between antibiotics and soil conditions, may retain antimicrobial properties, but have fates entirely different from the parent compound, complicating the task of assessing the persistence of a given drug and attached risks. Antibiotics detected in soils affected by agricultural practices are often in the μg/kg concentration range: 0.5 for trimethoprim, 1 for sulfadiazine, 2 for sulfamethazine, 8.5 for lincomycin, 11 for sulfadimidine, 39 for chlortetracycline; and up to 52 for ciprofloxacin, 67 for clarithromycin, 305 for oxytetracycline and 900 for tetracycline (Kemper, 2008). Antibiotics (*e.g.*, tetracyclines, fluoroquinolones, sulfonamides, amphenicols, trimethoprim) in the soil can be uptaken by plants (*e.g.*, carrot, celery, corn, lettuce, onion, potato) (Boxall, 2012); corn, lettuce and potato grown on manure-amended soil in a greenhouse had

0.1–1.2 mg/kg (dry weight) of sulfamethazine, with total uptake <0.1% of the amount added along with manure (Dolliver, 2007). Other reports put chlortetracycline concentrations in corn, green onion and cabbage tissues in the 2–17 μg/kg range (Kumar et al., 2005). Although concentrations found in plant tissues are "very low and unlikely to pose a risk to humans" (Boxall, 2012), they may exert effects upon pathogenic bacteria that also migrate into plant tissues: *E. coli* O157:H7 viable cells were recovered from inner tissues of lettuces grown in manure-contaminated soils (Solomon et al., 2002).

3.4 NON-ANTIBIOTIC SELECTIVE PRESSURES: HEAVY METALS, DISINFECTANTS, OTHER DRUGS AND BIOCIDES

Bacteria respond to numerous chemical insults by expressing unspecific defense mechanisms that can, in turn, protect them from antibiotics. Additionally, some specific resistance mechanisms, particularly those protecting bacteria from toxic concentrations of inorganic ions, usually reside along antibiotic resistance genes in the same mobile genetic elements, enabling for co-selection; mercury resistance genes have been extensively studied in this regard. Curiously enough, many metal ions are necessary for cell physiology and are only toxic at very high concentrations, not often found naturally; something that can be said also of naturally-occurring antibiotics which, at low concentrations, seem to have a physiological role for bacteria, being only toxic at the high, human-made concentrations. Other heavy metals, such as Ag, Cd or Hg, with not known biological role, are rather rare and natural exposure of bacteria would have been very limited; silver and mercury, however, have or had been used as disinfectants.

The anthropogenic release of heavy metals into the soil is massive, and increased dramatically since the 1950's. By year 2000, the cumulative global production of metals was 640,000 tons of Hg, 1.1 million tons of Cd, 36 million tons of Ni, 105 million tons of Cr, 235 million tons of Pb, 354 million tons of Zn, and 451 million tons of Cu (Han et al., 2002). Again, aquaculture and agriculture contribute importantly to the environmental release of some of these inorganic pollutants. Cu-containing compounds are used as anti-fouling agents in aquaculture cages and nets, and Zn, Cu, Cd and Pb are enriched in aquaculture sediments, while Hg is found in fish feed. Metals in fertilizers and sewage sludge applied to arable soil contribute to the release of Pb, Hg, Cd, Cu, Zn, Cr and Ni. Manure have been reported to contain (in mg/kg) up to 4 of Co, 6 of Cd, 28 of Ni, 40 of Cr, 100 of Pb, 220 of Cu, and 690 of Zn; up to 6 mg/kg of Hg have been found in sewage sludge. Consequently, dissolved metals in rivers have been detected to be in the tenths of μg/L in Asian rivers with agricultural and urban influence; but in sediments they reach up to hundreds of mg/kg (dry weight): 1.3 of Cd, 6 of Hg, 69 of Ni, 166 of Cu, 180 of Cr, 185 of Pb, and 1360 of Zn. Heavy metals are also detected in soils receiving compost or other fertilizers or pesticides, although at concentrations about a tenth of those of river sediments (Seiler and Berendonk, 2012).

Environmental pollution resulting in increased antibiotic resistance may come from unsuspected sources. Coal-fired power plants release coal ash containing several toxic metals that contaminate water bodies; in the Savannah river (South Carolina, US), such plants significantly increased the levels of arsenic (from 0.2 μg/L in a non-contaminated control, to 18.3 μg/L), strontium (from 22.5 to 186.8 μg/L), nickel (from 0.05 to 2.7 μg/L) and cadmium (from 0.02 to 0.13 μg/L), among others. Bacteria from the sites with higher metal levels did tolerate better experimental exposures to 1 mM cadmium or nickel and, surprisingly, to 300 µg/mL of tetracycline; this effect was much more clear in bacteria from the river sediment (Wright et al., 2006).

Drugs other than antibiotics are also released into the environment, especially from urban wastewater; some of these drugs have antibacterial activity: antihistamines, barbiturates, beta-blockers, diuretics, mucolytics, NSAIDs, proton-pump inhibitors, among others (Kristiansen, 1991). Concentrations at which they exert antibacterial activities are well above those find, if any, at wastewater; but the effects of exposure to sub-inhibitory effects of these compounds have not been well addressed so far.

Disinfectants can also contribute to the selection of antibiotic resistant bacteria, and are often released into wastewater. Quaternary ammonium compounds (QAC, e.g., benzalkonium chloride, cetrimide, tetraethylammonium bromide) are cationic detergents and disinfectants that are mostly used as fabric softeners, but that have detergent activity against biological membranes. The MIC ranges of benzalkonium (C_{12-16}) and dialkonium (C_{10}) chlorides are 25–700 and 4–250 µg/mL, respectively, being the lower value for *Listeria monocytogenes* and the higher for *P. aeruginosa*. A number of bacterial unspecific efflux pumps (*e.g.*, AcrAB-TolC, MexAB-OprM) confer resistance to quaternary ammonium compounds, as well as low-level antibiotic resistance (see Chapter 1); furthermore, specific resistance genes, such as *qacE*, are within the conserved region of class-1 integrons, allowing for co-selection by linkage. About 500,000 tons of QACs were consumed worldwide in 2004. Average concentrations of QACs are in the order of 500 µg/L in domestic wastewater, 50 µg/L in wastewater treatment effluents, 40 µg/L in surface waters receiving such effluents, and 5 g/kg (dry weight) of sewage sludge (Tezel and Pavlostathis, 2012).

Triclosan is a disinfectant commonly found in a variety of consumer products (soaps, detergents, even fabrics) as the flagship of the "aseptic" fade that pervaded households, especially in developed countries; while the FDA recently revoked the "Generally regarded as safe" status of triclosan, it is still present in many products worldwide. Triclosan and related compound triclocarban bind to enoyl-acyl carrier protein reductase, encoded by *fabI*, inhibiting in turn fatty acid synthesis; triclosan MICs ranged between 0.5 µg/mL for reference *E. coli* and *S. aureus* strains, to 64 µg/mL for clinical isolates (Assadian et al., 2011). Although other than a few over-expressing *fabI* mutants have been reported as reducing susceptibility to triclosan, the disinfectant is capable of selecting *mar* mutations that confer antibiotic resistance. At the influx of a wastewater treatment plant in the US (Baltimore), triclosan and triclocarban were detected at 6.1 and 6.7 µg/L, respectively. While treatment significatively reduced these concentrations (to 35 and 110 ng/L, respectively), they were still detectable in the effluent (Halden and Paull, 2005). On the other hand, triclosan and triclocarban captured from wastewater in sewage sludge is anyway released into the environment as such sludge is applied to land; an estimated 57 and 140 tons/year of triclosan and triclocarban are added to soils in the US (Halden, 2014).

There could be many more compounds with antibacterial activity that are released into the environment and that, in one way or another, can foster antibiotic resistance. Obvious candidates are other biocide compounds, such as herbicides, that are applied directly to soils reaching high concentrations. A single example, glyphosate, that modify the bacterial susceptibility to antibiotics (see next chapter), is applied in the amount of 100 million pounds per year, in the US alone, and the use of genetically-modified crops may dramatically increase this consumption rate. Demonstrated and hypothetical effects upon bacterial communities, of the release of antibiotics and other xenobiotics into the environment, will be discussed in further chapters.

REFERENCES

Assadian, O., Wehse, K., Hübner, N. O., Koburger, T., Bagel, S., Jethon, F. & Kramer, A. (2011) Minimum inhibitory (MIC) and minimum microbicidal concentration (MMC) of polihexanide and triclosan against antibiotic sensitive and resistant *Staphylococcus aureus* and *Escherichia coli* strains. *GMS Krankenhhyg. Interdiszip.*, 6, Doc06.

Boxall, A. B. A. (2012) Fate and transport of antibiotics in soil systems. In Keen, P. L. & Montforts, M. H. M. M. (Eds.) *Antimicrobial resistance in the environment*. Hoboken, John Wiley & Sons.

Collignon, P., Athukorala, P., Senanayake, S. & Khan, F. (2015) Antimicrobial resistance: the major contribution of poor governance and corruption to this growing problem. *PLoS One*, 10, e0116746.

Corcoran, E., Nellemann, C., Baker, E., Bos, R., Osborn, D. & Savelli, H. (2010) *Sick water? The central role of wastewater management in sustainable development*, Arendal, United Nations Environment Programme.

Dolliver, H. A. S. (2007) Fate and transport of veterinary antibiotics in the environment. University of Minnesota.

Edqvist, L. E. & Pedersen, K. B. (2001) Antimicrobials as growth promoters: resistance to common sense. In Harremoës, P., Gee, D., Macgarvin, M., Stirling, A., Keys, J., Wynne, B. & Vaz, S. G. (Eds.) *Late lessons from early warnings: the precautionary principle 1896–2000*. Copenhagen, European Environment Agency.

FAO, OIE & WHO (2006) *Antimicrobial use in aquaculture and antimicrobial resistance*, Seoul, FAO/OIE/WHO.

Graham, D. W., Olivares-Rieumont, S., Knapp, C. W., Lima, L., Werner, D. & Bowen, E. (2011) Antibiotic resistance gene abundances associated with waste discharges to the Almendares River near Havana, Cuba. *Environ. Sci. Technol.*, 45, 418–424.

Halden, R. & Paull, D. (2005) Co-occurrence of triclocarban and triclosan in U.S. water resources. *Environ. Sci. Technol.*, 39, 1420–1426.

Halden, R. U. (2014) On the need and speed of regulating triclosan and triclocarban in the United States. *Environ. Sci. Technol.*, 48, 3603–3611.

Han, F. X., Banin, A., Su, Y., Monts, D. L., Plodinec, J. M., Kingery, W. L. & Triplett, G. E. (2002) Industrial age anthropogenic inputs of heavy metals into the pedosphere. *Naturwissenschaften*, 89, 497–504.

Harms, K. & Bauer, J. (2012) Detection and occurrence of antibiotics and their metabolites in pig manure in Bavaria (Germany). In Keen, P. L. & Montforts, M. H. M. M. (Eds.) *Antimicrobial resistance in the environment*. Hoboken, John Wiley & Sons.

Huang, R., Ding, P., Huang, D. & Yang, F. (2015) Antibiotic pollution threatens public health in China. *Lancet*, 385, 773–774.

Johnson, A. C., Keller, V., Dumont, E. & Sumpter, J. P. (2015) Assessing the concentrations and risks of toxicity from the antibiotics ciprofloxacin, sulfamethoxazole, trimethoprim and erythromycin in European rivers. *Sci. Total Environ.*, 511, 747–755.

Keen, O. S. & Linden, K. G. (2013) Degradation of antibiotic activity during UV/H_2O_2 advanced oxidation and photolysis in wastewater effluent. *Environ. Sci. Technol.*, 47, 13020–13030.

Kemper, N. (2008) Veterinary antibiotics in the aquatic and terrestrial environment. *Ecol. Indicators*, 8, 1–13.

Khan, G. A., Berglund, B., Kahn, K. M., Lindgren, P. E. & Fick, J. (2013) Occurrence and abundance of antibiotics and resistance genes in rivers, canal and near drug formulation facilities – a study in Pakistan. *PLoS One*, 8, e62712.

Kristiansen, J. E. (1991) Antimicrobial activity of nonantibiotics. *ASM News*, 57, 135–139.

Kumar, K., Gupta, S. C., Baidoo, S. K., Chander, Y. & Rosen, C. J. (2005) Antibiotic uptake by plants from soil fertilized with animal manure. *J. Environ. Qual.*, 34, 2082–2085.

Kümmerer, K. (2003) Significance of antibiotics in the environment. *J. Antimicrob. Chemother.*, 52, 5–7.

Kümmerer, K. & Henninger, A. (2003) Promoting resistance by the emission of antibiotics from hospitals and households into effluent. *Clin. Microbiol. Infect.*, 9, 1203–1214.

Laxminarayan, R., Van Boeckel, T. & Teillant, A. (2015) The economic costs of withdrawing antimicrobial growth promoters from the livestock sector. *OECD Food, Agriculture and Fisheries Papers*, 78.

Levy, S. B. (2002) *The antibiotic paradox, 2nd ed.*, Cambridge MA, Perseus Publishing.

Moullan, N., Mouchiroud, L., Wang, X., Ryu, D., Williams, E. G., Mottis, A., Jovaisaite, V., Frochaux, M. V., Quiros, P. M., Deplancke, B., Houtkooper, R. H. & Auwerx, J. (2015) Tetracyclines disturb mitochondrial function across eukaryotic models: a call for caution in biomedical research. *Cell Reports*, 10, 1681–1691.

Pew Charitable Trusts (2014) *Gaps in FDA's antibiotics policy*, www.pewtrusts.org/en/research-and-analysis/issue-briefs/2014/11/gaps-in-fdas-antibiotics-policy.

Pruden, A., Larsson, D. G. J., Amézquita, A., Collignon, P., Brandt, K. K., Graham, D. W., Lazorchak, J. M., Suzuki, S., Silley, P., Snape, J. R., Topp, E., Zhang, T. & Zhu, Y. G. (2013) Management options for reducing the release of antibiotics and antibiotic resistance genes to the environment. *Environ. Health Perspect.*, 121, 878–885.

Segura, P. A., François, M., Gagnon, C. & Sauvé, S. (2009) Review of the occurrence of anti-infectives in contaminated wastewaters and natural and drinking waters. *Environ. Health Perspect.*, 117, 675–684.

Seiler, C. & Berendonk, T. U. (2012) Heavy metal driven co-selection of antibiotic resistance in soil and water bodies impacted by agriculture and aquaculture. *Front. Microbiol.*, 3, 399.

Singer, A. C., Järhult, J. D., Grabic, R., Khan, G. A., Lindberg, R. H., Fedorova, G., Fick, J., Bowes, M. J., Olsen, B. & Söderström, H. (2014) Intra- and inter-pandemic variations of antiviral, antibiotics and decongestants in wastewater treatment plants and receiving rivers. *PLoS One*, 9, e108621.

Solomon, E. B., Yaron, S. & Matthews, K. R. (2002) Transmission of *Escherichia coli* O157:H7 from contaminated manure and irrigation water to lettuce plant tissue and its subsequent internalization. *Appl. Environ. Microbiol.*, 68, 397–400.

Stockwell, V. O. & Duffy, B. (2012) Use of antibiotics in plant agriculture. *Rev. Sci. Tech. Off. Int. Epiz.*, 31, 199–210.

Suzuki, S. & Hoa, P. T. P. (2012) Distribution of quinolones, sulfonamides, tetracyclines in aquatic environment and antibiotic resistance in Indochina. *Front. Microbiol.*, 3, 67.

Tezel, U. & Pavlostathis, S. G. (2012) Role of quaternary ammonium compounds on antimicrobial resistance in the environment. In Keen, P. L. & Montforts, M. H. M. M. (Eds.) *Antimicrobial resistance in the environment*. Hoboken, John Wiley & Sons.

Tian, B., Fadhil, N. H., Powell, J. E., Kwong, W. K. & Moran, N. A. (2012) Long-term exposure to antibiotics has caused accumulation of resistance determinants in the gut microbiota of honeybees. *mBio*, 3, e00377–12.

Van Boeckel, T. P., Brower, C., Gilbert, M., Grenfell, B. T., Levin, S. A., Robinson, T. P., Teillant, A. & Laxminarayan, R. (2015) Global trends in antimicrobial use in food animals. *Proc. Natl. Acad. Sci. USA*, 112, 5649–5654.

Van Boeckel, T. P., Gandra, S., Ashok, A., Caudron, Q., Grenfell, B. T., Levin, S. A. & Laxminarayan, R. (2014) Global antibiotic consumption 2000 to 2010: an analysis of national pharmaceutical sales data. *Lancet Infect. Dis.*, 14, 742–750.

Wright, M. S., Peltier, G. L., Stepanauskas, R. & McArthur, J. V. (2006) Bacterial tolerances to metals and antibiotics in metal-contaminated and reference streams. *FEMS Microbiol. Ecol.*, 28, 293–302.

Zhang, X., Li, Y., Liu, B., Wang, J., Feng, C., Gao, M. & Wang, L. (2014) Prevalence of veterinary antibiotics and antibiotic-resistant *Escherichia coli* in the surface water of a livestock production region in northern China. *PLoS One*, 9, e111026.

Chapter 4

Spread of resistant organisms from human settlements into the environment

Human-made antibiotics, in huge amounts, reach the environment in a number of ways, as overviewed in the previous chapter. This alone is a cause of concern, as the ecological impact of the presence of these natural compounds in unnatural quantities, is hard to assess. However, the problem is worsened by the simultaneous release of resistant bacteria that have been selected by the human use of antibiotics. Again, this is happening at a multi-level, complex scenario that goes, from the seemingly insignificant open defecation of a pet dog receiving antibiotics in an urban setting; to the use of manure from medicated food animals to fertilize soils; to the massive release of untreated sewage into water bodies. These organisms can find their way into our foodstuff and drinking water, directly causing antibiotic-resistant infections; or can, possibly to a larger health risk, contribute their resistance determinants, and mobile genetic elements, to the enormous gene pool available via HGT, to be further selected by the concomitant presence of antibiotics in the environment.

Distinguishing resistance genes or resistant bacteria selected by the human use of antibiotics, from those already present in the environment, may be a tricky question. Some few reports can clearly make that distinction; for instance: (a) DNA samples from 30,000-years-old permafrost, most likely free of recent inputs (D'Costa et al., 2011); and (b) a culture-based study of a cave system assumedly isolated from surface input for the last 4–7 million years (yet another paper with arbitrarily set resistance breakpoints (a 20-µg/mL concentration for all antibiotics tested) and incapable of distinguishing intrinsic resistance phenotypes; (Bhullar et al., 2012)). When DNA sequences of resistance genes found in the environment are available, phylogenetic analysis can prove that some of those genes are likely apart from the typical ones found in clinical isolates (e.g., (Voolaid et al., 2013)). But in most other cases, it is difficult to assess whether a resistance gene or resistant bacteria found in the environment, was selected by the use of antibiotics, clinical or otherwise, and then released to the environment; or if it was selected from within environmental bacteria by human-made antibiotics or other pressures, human-related or not. Even studies attempting to measure the impact of human activities, e.g., comparisons between aquatic isolates up- and down-stream of wastewater treatment plant effluents, or establishing a gradient of farming activities, can be affected by some uncertainty.

Many studies on the purported release of resistance into the environment, as opposed to ancient resistance unrelated to the human use of antibiotics, are centered on well-known bacteria (mostly E. coli) or genes (mostly those encoding ESBLs). The vast availability of laboratory or bioinformatic tools and knowledge on these two enables for easy surveillance and result interpretation. Also, the particular combination of ESBL-*E. coli*, which was first detected in the clinical setting, is considered to be more likely originated from human influence (as such, or as the result of the acquisition of ESBL genes from human settlements, by indigenous *E. coli*), than something that arise on their own; although this is essentially an speculation. Anyhow, by only focusing on this pair, we are most certainly missing a bigger picture. On the other hand, wider studies trying to assess, either by culture- or molecular-based methods, the extent of resistance in other bacterial species, and to antibiotics other than beta-lactams, may be plagued by the same problems that affect most other environmental resistance surveillances: lack of adequate resistance breakpoints (or even of functional definitions of resistance), confusion between intrinsic, acquired and adaptive resistance (and the relevance of each one), ill-curated resistance gene databases, etc. Conclusions drawn from these studies, while attempting to look at said bigger picture, may get entangled with the many unknowns of this field. The following sections of this chapter will review some of the evidence regarding the release of resistance into the environment.

4.1 CLINICAL ENVIRONMENTS TO URBAN ENVIRONMENTS TO RURAL ENVIRONMENTS

Without getting much into documenting these data, each human produces about 120 g of feces per day; considering that about half of the current human population, *i.e.*, ~3.5 billion people now live in cities, about 420,000,000 kg of feces are released into wastewater each day (wrongfully assuming that all people living in cities have access to sewerage). Bacteria comprise up to half of the total solids in feces (Stephen and Cummings, 1980). All sorts of antibiotic resistance genes and mobile elements have been detected by metagenomic analyses of the human intestinal microbiota (Broaders et al., 2013). Most importantly, at population levels, the presence of resistance genes is not dependent on the individual usage of antibiotics but rather on their agricultural use and age, and vary from country to country (Forslund et al., 2013); therefore, antibiotic resistant bacteria are not only released by antibiotic-receiving people, but actually by the whole urban population. Urine is not often considered as a source of bacteria, as it is (wrongfully) supposed to be sterile, but for people suffering urinary infections or asymptomatic bacteriuria; however, a single person with such a condition can be releasing 50–250 millions of uropathogenic, often multi-resistant bacteria into urban wastewater daily. By simply extrapolating the yearly 7 million medical office visits due to urinary infections in the US (Foxman, 2003) to the global urban population, up to 75 million people could be infected each year; each day these sick people urinate, they could be releasing nearly 2×10^{16} uropathogenic bacteria into the sewage. Wastewater carry A LOT of resistant bacteria. Needless to say, the other half of the human population, *i.e.*, the one that does not live in cities, release excreta into small-scale septic tanks or directly into the environment.

Human-to-human transmission of resistant bacteria, either pathogenic or innocuous, is not within the purview of this book, as it is in the realm of clinical microbiology. However, it is clear that humans and animals, especially those receiving antibiotics, shed resistant bacteria; the impact of this shedding in urban environments is mostly unknown, with much more evidence referring to farm animals and their waste. Fecal pollution in urban settings come from a variety of sources: open-air fecalism from humans, pets, stray dogs and cats, and other animals, along with sewer leaks, failing septic tanks, and improper disposal of garbage (which often contains used disposable diapers and pets' excreta). While these situations are common in non-developed countries, they also affect developed ones: it is estimated that dogs leave 82,000 kg of feces on the ground in the US alone (and Germany and some US cities are trying now a quaint anti-urination painting strategy for some urban areas). During dry seasons, bacteria from these sources can get airborne; in Mexico City, children can ingest up to 200 fecal coliforms (and 75 bacterial endotoxin units) per day, and adults about the double, coming from dust and airborne particle matter (Rosas et al., 2011). In a report from the same city, there were important differences between the number of *E. coli* isolates, and their antibiotic resistance, comparing indoor and outdoor samples. There was an average of twice as more *E. coli* cells (most probable number, MPN) per gram (1089) of indoor dust (especially if houses had carpets or pets) than in outdoor dust (435); however, resistance to at least one antibiotic was more frequent outdoors (73% *vs.* 45%). Resistance to ciprofloxacin, chloramphenicol, co-trimoxazole, and tetracycline were higher in outdoor isolates (resistance to ciprofloxacin was *only* found in outdoor isolates, at rates of 27–36%); while resistance to ticarcillin/clavulanate (at rates, 15–22%, almost identical to ampicillin resistance, suggesting a chromosomal beta-lactamase) were only found in indoor isolates (also suggesting that the 21–25% ampicillin resistance rate outdoors was likely plasmid-mediated) (Rosas et al., 2006). It is possible that the increased environmental pressures outdoors diminish the number of viable *E. coli* cells, but that remaining organisms tend to be more resistant; something akin to what is known to happen in wastewater treatment plants, as will be discussed below. While entertaining this notion, it is also possible to surmise that such urban outdoor environmental pressures also favor the presence of mobile elements: plasmids as suggested by the evidence above; or class-1 integrons, as measured upon the same *E. coli* isolates: up to 15% of outdoor strains carry such elements, but only 2% of indoor isolates do (Díaz-Mejía et al., 2008). Bacteria from urban soils (again Mexico City) were frequently resistant to lead (MICs of 800–1600 μg/mL), and often also resistant to arsenate, chromate, cadmium and mercury (Vaca-Pacheco et al., 1995). Heavy metals may play a role in co-selecting antibiotic resistant bacteria; this will be briefly discussed at the end of this chapter.

A particularly disturbing report of resistance in urban settings was the finding of diverse isolates producing the New Delhi Metallo-beta-lactamase, NDM-1 (a carbapenemase conferring resistance to most beta-lactams, first reported in Sweden, in a *K. pneumoniae* isolate from a patient of Indian origin, and then spread worldwide at an amazing pace), in waste seepage and even tap water in New Delhi. The encoding gene was found mostly in plasmids (ranging 140–400 kb), from several *Pseudomonas* spp., *E. coli*, *K. pneumoniae*, *Citrobacter freundii*, *Shigella boydii*, *Aeromonas caviae*, *Vibrio cholerae*, *Kingella denitrificans* and *Achromobacter* spp.; in nearly half of the isolates, the gene did not actually confer carbapenem resistance. Importantly, only 60% of the New Delhi population has access to the sewerage system (Walsh et al., 2011). But this is hardly a picture exclusive to non-developed countries: in the German city of Mainz, the *vanA* gene was detected by PCR, in the absence of enterococci, in drinking water biofilms (Schwartz et al., 2003).

The role of antibiotic usage "hotspots" in cities, *i.e.*, health-care facilities, as sources of resistant bacteria later found in urban environments has not been addressed often. Bacteria under the extreme selective pressures within such hotspots, are released through wastewater and garbage:

- *Acinetobacter* isolates from hospital sewage in Denmark, for instance, were often resistant to chloramphenicol (up to 55%) and oxytetracycline (up to 37%); although much less resistant than those found at pharmaceutical plant sewers: up to 61% resistance to both antibiotics mentioned before, and 58% to sulfonamides, 21% to gentamicin and 31% to amoxicillin (38.6% resistant to 3 or more antibiotics, *vs.* 4.7% in hospital sewage; (Guardabassi et al., 1998)).
- Enterococci isolated from urban sewage, up- and down-stream hospitals in Portugal, went from 0 to 27% resistant to vancomycin, 11 to 45% resistant to ampicillin, and 27 to 75% resistant to ciprofloxacin (while resistance to nitrofurantoin decreased from 14 to 6%); this wastewater is drained to a river estuary, where resistance rates were 33, 33 and 56%, respectively (Novais et al., 2005).
- Resistant and multi-resistant *P. aeruginosa* were significantly more frequently isolated from hospital wastewater than from general urban wastewater in Besançon, France; authors of this study highlighted the fact that, while dilution in the community is only in the order of 150 L/inhabitant·day, it is about 1,000 L/bed·day in hospitals, which should reduce the presence of pathogenic bacteria in wastewater (Slekovec et al., 2012).
- *mecA* gene was detected by PCR in wastewater biofilms from German hospitals, while it was not detected in municipal sewage, wastewater treatment plant or river surface water (Schwartz et al., 2003).

Bacteria also travel on and in people leaving the hospital: discharged patients, relatives and other visitors, and healthcare personnel–it is so common, around the large hospitals in Mexico City, to see the young training physicians arriving or leaving in coats and scrubs, most likely covered with the hospital microbiota. Other vehicles for resistant bacteria can be unsuspected: pet cats, that are common at veterinary offices and hospitals, carry enterococci that are very often (49%) resistant to at least 3 antibiotics, mostly tetracycline (75%), erythromycin (50%) and rifampicin (36%), but even vancomycin resistant strains were found in one of six cats sampled (Ghosh et al., 2012). Cats tend to roam and defecate outdoors. Raccoons roaming at the Toronto zoo carried a larger amount and variety of resistant *E. coli*, including ESBL-producers (Jardine et al., 2012). Human hospitals are thought of as clean, vermin-free locations, but the reality in non-developed countries is far from it: flies, cockroaches and rats are common, and even stray dogs roam within some. Hospital cockroaches have been found to carry nosocomial *Acinetobacter* (Casellas and Quinteros, 2007) and *Klebsiella* (Cotton et al., 2000) strains, likely contributing to hospital outbreaks, but also to the spread of these organisms outside hospital walls and into the urban environment. Routes, varieties, resistances and amounts of hospital bacteria leaking into the urban environment are mostly unknown.

Information on the nature and scale of the urban microbiota is scarce (King, 2014); and on the urban antibiotic resistome is nearly null. Also, it is possible to speculate on the differences between developed and non-developed countries, with the latter

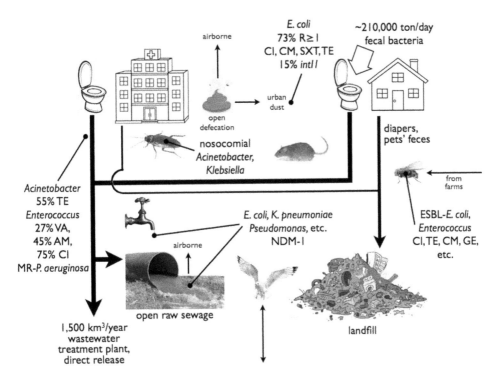

Figure 4.1 **Resistant bacteria in the urban environment.** The nature and extent of antibiotic resistance in the urban environment is mostly unknown. Assumedly, resistant bacteria and resistance genes derive from the clinical use of antibiotics, within hospitals and by outpatients. Most abundant sources would be excreta (feces and urine), which would end up in the sewerage; this would leave the city, unless raw sewage gets exposed, by deficient infrastructure, leaks, floods, etc. Additionally, open defecation adds to the urban microbiota, particularly in non-developed countries. Dried feces, particularly in dry seasons, contribute to urban dust, which gets airborne, potentially spreading resistant bacteria within and beyond the urban setting. Fecal material from houses, and contaminated material from hospitals, may end up in landfills, accessible to urban animals (rats, flies, cockroaches, stray dogs) and even people. Animals may play an important role in mobilizing resistant bacteria, especially airborne ones: birds are known to carry many different resistant organisms; flies import resistant bacteria from farms into cities; and hospital cockroaches carry nosocomial, multi-resistant pathogens. Some examples are indicated (see text for further details): most *E. coli* in urban dust are resistant to at least one drug; resistant *Acinetobacter* and enterococci were found in the sewage of hospitals; ESLB-producing *E. coli* and multi-resistant enterococci are carried by flies from farms and wastewater treatment plants; diverse bacteria carrying genes for the New Delhi Metallo-beta-lactamase (NDM-1) were found even in tap water. AM, ampicillin; Cl, ciprofloxacin; CM, chloramphenicol; GE, gentamicin; SXT, sulfamethoxazole-trimethoprim; TE, tetracycline; VA, vancomycin.

being likely to include more human or animal resistant pathogens, and more selective and maintenance pressures, in the direct form of antibiotics, or as other chemical agents favoring resistant bacteria by co-selection. Many of the microbes in urban soil and dust, end up in rural environments, carried by water (sewage, runoff), wind or airborne animals; the extent of such dispersal is completely unknown. Pathogens deposited on urban ground may interact with soil bacteria, exchanging resistance and mobility determinants; all occurring under the selective pressures particular to urban environments. But all these belong, so far, to the realm of speculation.

4.1.1   Urban wastewater

As reviewed in the previous chapter, most wastewater is released into the environment without previous treatment, as treatment plants are mostly a feature of developed countries. While 90% of wastewater in US and Canada is treated, and 66% in Europe, only 35% in Asia, 14% in Latin America and less than 1% in Africa is. The daily production of wastewater, globally, was ~1,500 km^3 by year 2000. In this way, antibiotics, in the ng/L to µg/L concentrations (see previous chapter), are directly released along with sewage bacteria, which are mostly human commensals (and some pathogens), into receiving water bodies. Soils are also affected: Mexican wastewater not undergoing treatment, for instance, is also used for irrigation purposes. In a study of a valley that have been receiving raw wastewater (65% domestic sewage, 20% from service sector and 15% industrial) for different time spans, genes related to class-1 integrons (*intI1*, *qacE* and *aadA*) and IncP-1 plasmids (*korB*) increased steadily and seem to accumulate over time, up to 10^6 gene copies per gram of dry soils that have received such water for 85 years or more (Jechalke et al., 2015). But before analyzing the effects of the direct release of wastewater into the environment, lets overview what happens in treatment plants; apparently, it is fortunate that only a minority of global wastewater undergo treatment, at least as antibiotic resistance is concerned.

4.1.1.1   Wastewater treatment plants

The word "treatment", referring to wastewater, includes a wide variety of processes: a primary phase aimed at reducing total suspended solids, often include sedimentation and septic tanks, and some primary chemical treatment; a secondary phase reduces the biochemical and chemical oxygen demand, in aerated lagoons, activated sludges, filters and digesters; and a tertiary phase further reduces oxygen demand as well as nitrogen, phosphorous, and microbial removal, using filtration and/or disinfection. Treatment plants around the world differ in the nature of their processes, so it is not really adequate to refer to "treatment" as a standard, uniform process worldwide. However, for the purposes of this book, and due to the lack of specific information regarding the impact of each type of process upon resistant bacteria, the following paragraphs would only refer to "wastewater treatment", without further specifics.

The role of wastewater treatment plants on the enrichment and dispersal or antibiotic resistant bacteria has been extensively studied. Treatment reduces bacterial load by several orders of magnitude; however, the notion that survivors are more resistant to antibiotics has been entertained for a while. Chlorination, in particular, has been thought to select for antibiotic resistance (Murray et al., 1984). *Acinetobacter* spp., for instance, went from 9% amoxicillin/clavulanate resistance in raw sewage, to 38% after treatment; similar increases were noticed for chloramphenicol (from 25% to 69%) and rifampin (from 63% to 84%), and in multi-resistance (defined as resistance to three or more: from 33% to 72%) (Zhang et al., 2009). A study following the path of specific resistance (*sul1* and *sul2*, mediating sulfonamide resistance) genes along a wastewater path, from hospital to wastewater treatment plant to effluent to water body, done in Switzerland (Lausanne), showed the decline in *sul* resistance gene presence, from $\sim 10^6$–10^7 copies/mL in hospital wastewater to 10^5–10^6 copies/mL in plant effluent, to 10^2–10^3 copies/mL in the lake (corresponding with sulfamethoxazole concentrations of 1,116, 61, and 0.4 ng/L, respectively). However, the proportion of *sul1*/16S-rRNA copy numbers increased through treatment, from $\sim 3\%$ at the influent, to $\sim 8\%$ at the effluent; as did the rate of isolates resistant to 8 or more antibiotics, from $\sim 30\%$ to $\sim 60\%$, respectively (Czekalski et al., 2012). Treatment plants' diverse microenvironments are therefore considered as a sort of funnel, sites where resistant bacteria, mobility genes, and selective pressures, arrive and interact; HTG is also enhanced, by the close proximity and adequate conditions provided by sludge (see below). A PCR-based study upon sludge bacteria (pre-cultured on antibiotic-containing media) from a German wastewater treatment plant, revealed the presence of all sorts of resistance genes: 24 aminoglycoside-modifying enzymes, 33 beta-lactamases (including metallo-enzymes), 18 chloramphenicol-resistance genes (acetyltransferases and efflux pumps), 4 *qnr* genes, 14 macrolide-resistance genes (modifying enzymes and efflux pumps), 14 sulfonamide and trimethoprim resistance genes, 22 *tet* genes, 5 quaternary-ammonium compound resistance genes, as well as genes specific for plasmids of incompatibility groups N, Q, A/C, FIA and P (Szczepanowski et al., 2009). Very complex multi-resistance plasmids, likely to have been assembled through stepwise integration, transposition and recombination processes, have been found in bacteria from wastewater treatment plants; an IncF, 120-kb conjugative plasmid conferring resistance to ampicillin, chloramphenicol, erythromycin, kanamycin, neomycin, streptomycin, sulfonamides, trimethoprim and tetracycline, as well as to mercury, containing complete or remnant regions of transposons Tn*21*, Tn*10*, Tn*1*, and Tn*402*; insertion sequences IS*26* and IS*6100*, and a class-1 integron (Szczepanowski et al., 2005), is a nice example of the kind of mobile elements found in wastewater treatment plants.

A formidable analysis (Schlüter et al., 2007) on the nature of IncP-1 plasmids isolated from sewage sludge bacteria at wastewater treatment plants, provide evidence for the extensive intra- and inter-cellular mobilization of genes in such environments. The arrival of resistant bacteria, mainly from medicated people at houses and hospitals, into such plants, allow or even promote genetic exchange: high bacterial densities, attached to suspended organic material or forming biofilms, under the selective pressure (or HGT-inducing activities) exerted by antibiotics, disinfectants, detergents, heavy metals and other agents, make treatment plants into resistance and mobility boosters; and IncP-1 plasmids into the powerhouses of accumulation and spread of resistance genes. And while the load of resistant bacteria may be reduced in the water column due to treatment in wastewater plants (up to 99.8% according to one report (Yang et al., 2014)), sludge retains most of these bacteria; further disposal of such sludge, some times in landfills or even as fertilizers or soil conditioners, releases huge amounts of resistant bacteria into the environment. Summarizing in a single phrase, "conventional wastewater treatment (*i.e.*, activated sludge) tends to make bacterial populations in treated effluent more frequently resistant, resistant to a larger variety of antimicrobials, and more likely to be resistant to multiple antimicrobials compared to those in raw influent" (LaPara and Burch, 2012).

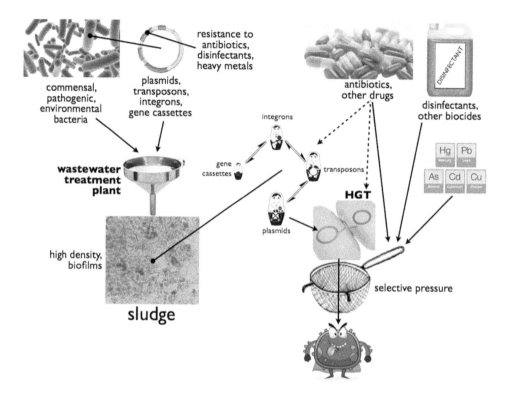

Figure 4.2 **Wastewater treatment plant sludge: brewing mobile multi-resistance.** Urban wastewater receives huge amounts of commensal and pathogenic bacteria, many carrying antibiotic resistance genes in mobile genetic elements. As a consequence of first and second stages of treatment, organic material, along with most bacterial load, is sedimented into a sludge; this puts bacteria in close proximity, allowing for HGT of all sorts. This happens under the effects of antibiotics (and their metabolites), other drugs with antimicrobial properties, disinfectants, and heavy metals, among others. These effects could be "simply" a selective pressure, favoring the resistance phenotype and potentially co-selecting for other determinants; or could be a subtle influence, modifying gene expression or mobility. The result of this complex combination is often a sum of gene rearrangements ending up in highly multi-resistant, genetically versatile organisms, which are then released to waters and soils.

4.1.1.2 *After treatment: release into water bodies*

Finding antibiotic resistance in freshwater is not new. An old paper (1978) reports detecting 17–24% of tetracycline resistance, 14–18% of ampicillin resistance, 4–17% of neomycin resistance, and even 2% of nalidixic acid resistance, among fecal coliforms isolated from rivers and a bay at Oregon (US) (Kelch and Lee, 1978); curiously, nearly all were resistant to nitrofurazone–resistance to nitrofurans among *E. coli* is rare even in clinical isolates nowadays suggesting, once again, the pervasive presence of intrinsic resistance within these data. Much more recently, a functional metagenomic analysis of water samples collected in a river both, upstream and downstream of a treatment plant, revealed a significant difference in the number of clones that were resistant to gentamicin, neomycin, amikacin and ciprofloxacin, being obviously more frequent when receiving DNA from downstream samples. Along with known beta-lactamases and aminoglycoside-modifying enzymes, other, previously unknown resistance determinants were found in clones from downstream samples (Amos et al., 2014), suggesting that many of these "resistance" genes came along with urban wastewater, either from humans or animals, or the sewage environment. In the same setting, the prevalence of class 1 integrons detected by quantitative PCR, also revealed the significant input of wastewater treatment plants, and allowed for modeling capable of predicting the influence of other variables in the magnitude of such input (Amos et al., 2015). Surprisingly, an Australian study of antibiotic resistant *E. coli* isolated from a wastewater treatment plant, and of downstream surface waters and oysters, showed lower resistance rates in isolates from the plant itself, except for ampicillin and nalidixic acid: from 10% to 51% for tetracycline, from 17% to 32% for sulfonamide, from non detectable to 8% for gentamicin; even resistance to $\geq$1 antibiotic was lower in plant isolates (from 31% to 87%) (Watkinson et al., 2007).

An urban lake in Mexico City is an interesting example of the many conditions of non-developed countries interacting. Until very recently (2011), antibiotics were sold without medical prescription in Mexico, and antibiotic usage, in DDD/1000-inhab/day ($\sim$14–16), was the higher in Latin America up to 2005 (Wirtz et al., 2010). This led to the highest resistance rates among many pathogens, within Latin America, or even worldwide (Amábile-Cuevas, 2010). Considering the relationship between resistance and corruption (Collignon et al., 2015), and Mexico ranking 103 out of 175 in Corruption Perception Index, high resistance is no surprise. While it is formally reported that the agricultural use of antibiotics was "banned" in Mexico since 2007, many relevant drugs (avoparcin, vancomycin, virginiamycin, spiramycin, etc.) are still allowed for growth promotion (Maron et al., 2013); and there is no real enforcement for the ban, so abuse is rife. While the actual "Federal District" of Mexico has only about 9 million inhabitants, the whole metropolitan area of Mexico City has 20.4 million inhabitants, making it the sixth most inhabited city in the world. Nevertheless, only $\sim$10% of wastewater undergo treatment, some of it only partially (lacking chlorination steps). This treated and semi-treated water is used to refill an urban lake, Xochimilco, which is a touristic attraction due to the pre-Columbian era agriculture system in floating garden plots (*chinampas*); the wastewater of surrounding houses often goes directly into the lake as well. By measuring resistance along the way, some interesting information can be obtained. Samples taken from the treatment plant effluent have averages of nearly 70 million bacteria per mL, 2 millions of them total coliforms, and 135,000 fecal coliforms; they are mostly of human origin, according to a non-too-reliable fecal coliform/fecal enterococci rate. *E. coli* isolates are resistant to tetracycline (32%), ampicillin (20%, most susceptible to ampicillin/sulbactam), co-trimoxazole (13%) chloramphenicol (11%), and ciprofloxacin (8%); while $\sim$50% lower, these resistance rate profile is similar to the one of enteropathogenic *E. coli* clinical isolates from the same city. Once in the lake, bacterial load is greatly diminished, but it does so in a differential way: total bacteria, total coliforms and fecal coliforms are only 0.02%, 0.001% and 0.008% of treatment plant figures, respectively, in the water column; but 6.6%, 2.7% and 3.5%, respectively, in the sediment, indicating that *E. coli* tend to accumulate there (posing an additional health risk: sediments are routinely removed and used as fertilizer). Resistance is also diminished in differential ways, with bacteria isolated from the water column having only about half of the resistance rates indicated above, but those from the sediment having 60–75% of the values observed in plant effluent isolates (Rosas et al., 2015). Resistance to ciprofloxacin and chloramphenicol seem to have some differences. Ciprofloxacin resistance is much lower in the lake than in the plant effluent (<1% in the water column, 3% in the sediment), and resistant isolates from the lake have a resistance profile significantly different from clinical, ciprofloxacin-resistant isolates: higher tetracycline and co-trimoxazole resistance, and lower ampicillin and ceftazidime resistance. Furthermore, ciprofloxacin MICs are significantly lower in lake resistant isolates (mode 8 μg/mL) than in clinical isolates (mode 256 μg/mL); bacteria with MICs of 4–16 μg/mL are almost non existent among clinical isolates, but make up to 30% of environmental ones. This suggests that an additional input of low-level ciprofloxacin resistant bacteria exists in the environment, perhaps caused by low-level exposure to fluoroquinolones (concentrations achieved in urine and feces of medicated patients can reach hundreds of μg/mL), or other selective pressures in the environment (Amábile-Cuevas et al., 2009). As to chloramphenicol, a somewhat similar phenomenon occurs: there were actually more resistant *E. coli* isolates in the lake sediment (13%) than in the plant effluent (11%), but MICs were higher in the plant effluent (average 308 μg/mL) than at the lake water column (average 130 μg/mL) or sediment (average 214 μg/mL), while all chloramphenicol-resistant isolates carried *catI* genes both, in plasmids and chromosomes (Rosas et al., 2015). The overall conclusions to be drawn from this system are that treated water still carry many resistant fecal bacteria, which get diluted while arriving to a new water body, and distribute differentially, with most *E. coli* going to the sediment; other sources of enteric bacteria and/or other selective pressures contribute with lower-level resistance to some antibiotics (ciprofloxacin and chloramphenicol), rapidly modifying the profile of bacteria released from the treatment plant. A final recommendation for tourists visiting Xochimilco: do NOT drink or dip in the water of the lake.

There are far too many reports of resistance found in water bodies all over the world: urban and rural; up- and down-stream of hospitals, cities, wastewater treatment plants, pharmaceutical companies or farming sites; at developed and non-developed countries. Selected examples are summarized in Table 4.1; this is not intended to be an exhaustive review on the issue, but only to provide a notion of the diversity of resistant bacteria and resistance genes out there–as well as of the diversity of approaches. Each paper provides an interesting edge; but from the raw data it evident that (a) rivers all over the world carry a variety of antibiotic resistant bacteria; (b) while resistant to old antibiotics (*i.e.*, sulfonamides, tetracyclines, aminopenicillins) are more common, resistance to newer ones (*e.g.*, fluoroquinolones, third-generation cephalosporins) and to those mostly confined to hospital use (*e.g.*, aminoglycosides, carbapenems) can also be found; (c) human presence and, most significantly, wastewater treatment plants, exert a clear influence on the prevalence of resistant bacteria or resistance genes found in water bodies, but (d) this influence differs between antibiotics; and (e) there is no clear difference between developed and non-developed countries. The good news is that antibiotic resistance genes do not seem to have reached deep sea sediments (Chen et al., 2013).

Table 4.1 Examples of resistance surveillance studies on water bodies affected by human activity.

Site	Source type	Microorganism	Genotype or phenotype[a]
non-developed countries			
Cameroon (New Bell, Douala) (Akoachere et al., 2013)	streams and wells	V. cholerae	92% multi-resistant (AM, TE, CM, SXT)
India (Mahananda River, West Bengal) (Chakraborty et al., 2013)	River	Comamonas, Acidovorax, Moraxellaceae, Pseudomonadaceae, Aeromonadaceae, Enterobacteriaceae	76% resistant to at least 1 antibiotic; 92/2188 carrying class-1 integron
China (wwtp, river) (Li et al., 2010)	tetracycline-eliminating wwtp, and up-, down-stream	Gammaproteobacteria (mostly Pseudomonas, Stenotrophomonas)	up-/wwtp/down-stream: CT 2/71/60%, TE 3/95/86%, CI 8/9/7%, CM 9/73/83%, AM 25/85/93%, RF 26/88/92%; number of resistances (mode): 1/6/6; integron carriage, from 95% without, to 30% with 3, to 26% with one.
China (Jiyun River, Pinggu County) (Zhang et al., 2014)	river surrounded by livestock production	218 E. coli isolates	CM 18%, LE 30%, SXT 40%, GE 45%, AM 50%, TE 50-60%
China (Jinxi River) (Lu et al., 2010)	urban river, 1 km downstream hospital	ESBL producers, 39/56 Enterobacteriaceae	IM 2%, CI 20%, GE 62%, TE 72%, SXT 73%
Mexico (San Pedro River) (Ramírez Castillo et al., 2013)	river receiving urban, industrial and livestock farm wastewaters	E. coli (10^3/100 mL total coliforms, 2.5×10^2/100 mL fecal coliforms)	CT 1%, GE 2%, LE 4%, NF 6%, CM 22%, AM 27%, SXT 39%
developed countries			
Portugal (diverse rivers) (Tacão et al., 2012)	polluted and unpolluted rivers	MacConkey-culturable; ESBL producers, mostly E. coli, Aeromonas hydrophila, Pseudomonas spp.	growing on 8-μg/mL CT Mac, 8.8% of culturable from polluted rivers, 0.6% from nonpolluted; 17% CT-resistant, ESBL producers, 18/39 bla_{CTX-M}, linked to ISEcp1

(Continued)

Table 4.1 Continued.

Site	Source type	Microorganism	Genotype or phenotype[a]
Spain (Ter River) (Marti et al., 2013a)	up-, wwtp, down-stream	bacteria from biofilms and sediments	$qnrA$, $qnrB$, undetected upstream; ~1 order of magnitude fewer copies per 16S rRNA in biofilms upstream: $tet(O)$ 10^{-5}, $tet(W)$ 10^{-5}, bla_{SHV} 10^{-5}, $qnrS$ 10^{-4}, bla_{TEM} 10^{-4}, $erm(B)$ 10^{-4}, $suIII$ 10^{-4}, $suII$ 10^{-3}; similar levels bla_{CTX-M}, and in sediment of all samples
US (14 rivers) (Ash et al., 2002)		Acinetobacter, Alcaligenes, Citrobacter, Enterobacter, Pseudomonas, Serratia	up to CT 9%, IM 10%,AMC 68% AM 73%
US (7 rivers) (Aubron et al., 2005)		Enterobacter asburiae	IMI-2 beta-lactamase in conjugative plasmids, 99% identity to E. cloacae chromosomal IMI-1
UK (Thames River) (Dhanji et al., 2011)		E. coli	CI-resistance in ST131 clone; $bla_{CTX-M-14}$ in some, but no $bla_{CTX-M-15}$ (most common in clinical isolates)
Switzerland (rivers and lakes) (Zurfluh et al., 2013)			21/58 water bodies positive for ESBL; one for carbapenemase-producing K. pneumoniae; 71/74 isolates bla_{CTX-M}, 3/74 bla_{SHV}
Switzerland (rivers and lakes) (Zurfluh et al., 2014)		ESBL-producing E. coli	70% also quinolone resistant, 40/42 QRDR mutations, 4 QnrS1, 15 AAC-6'-Ib-cr, 1 QepA
France (Seine estuary stations) (Laroche et al., 2009)		E. coli (36-4300 CFU/100 mL)	30–56% resistant to ≥1; CT 0–4%, CI,AK 0–10%, GE 0–12%, SXT 5–17% AMC 8–15% CM 8–38%, AX 11–29% TE 22–44% $intI$ 8.9%, $intI2$ 1.4%, both 0.7%
France (Leclercq et al., 2013)	medical centers to wwtp to river	enterococci (89–98% E. faecium)	medical center to wwtp plume: TE 21% to 24%, MC 87% to 72%, PE 100% to 4%, FQ 100% to 5%

a. Percentage figures are of resistant bacteria.
AM, ampicillin; AMC, amoxicillin-clavulanate; AK, amikacin; AX amoxicillin; CI, ciprofloxacin; CM, chloramphenicol; CT, cefotaxime; GE, gentamicin; IM, imipenem; LE, levofloxacin; MC, macrolide; NF, nitrofurantoin; PE, penicillin; RF, rifampicin; SXT, sulfamethoxazole-trimethoprim; TE, tetracycline; wwtp, wastewater treatment plant.

4.1.2 Resistance in soils

Resistant bacteria from contaminated water bodies, as well as from the application of sludge from wastewater treatment plants, and of manure and other residues from farms, also end up in the soil. As reported earlier in this chapter, the use of wastewater for irrigation purposes increase the load of mobility genes in soils; but this is hardly the only detectable change: microbial biomass and activity is increased by wastewater irrigation, as well as shifts in the composition of microbial communities. However, antibiotic resistance genes were actually more frequently found in soils irrigated with freshwater than in those receiving wastewater, leading the authors to conclude that resistant bacteria from wastewater do not survive for long in the soil, and that antibiotics in wastewater do not exert a significant selective pressure upon the native soil microbiota. Other comparative analysis of soils revealed that isolates from those receiving wastewater were more frequently resistant to daptomycin and lincomycin, but those receiving groundwater were more resistant to erythromycin, tetracycline and ciprofloxacin, and were more commonly multi-resistant (Gatica and Cytryn, 2013).

An entirely different picture results from the application of manure. An example of the bacterial load of swine and dairy manure was recently reported: $\log_{10}$ CFU/g (wet weight) were around 5 for fecal coliforms and enterococci in both types of manure; and other species (*Clostridium perfringens*, *Yersinia*, *Campylobacter* and *Aeromonas* spp.) in the 2.6–6.4 range. Animals in this sample routinely received penicillin, tetracyclines and sufamethazine; isolates from swine and dairy manure were resistant to ampicillin (18% and 35%, respectively), amoxicillin-clavulanate (1% and 11%), chloramphenicol (0.1% and 1%), ciprofloxacin (1% and below detection), gentamicin (0.3% for both), sulfamethoxazole (11% and 31%), tetracycline (5% and 32%), and sulfamethoxazole-trimethoprim (11% and 19%) (Marti et al., 2013b). A metagenomic library from cow manure was functionally tested for beta-lactam, chloramphenicol, tetracyclines and aminoglycosides, finding 80 different resistance genes, with 50–60% identity to previously reported sequences (Wichmann et al., 2014). To further aggravate this picture, manure of medicated animals also contains high concentrations of antibiotics, as was reviewed in the previous chapter.

As with reports of resistance in wastewater and receiving water bodies, the presence of resistance in the soil after the application of manure has been extensively reported; likewise, the surveillance approach, and the nature of resistance genes or resistant bacteria detected varies widely. For instance, resistance to tetracycline (Agersø and Sandvang, 2005, Kyselková et al., 2015) sulfonamides (Heuer et al., 2011, Wang et al., 2014) or the presence of ESBLs (Gao et al., 2015) in manured soils, has been reported. A quantitative PCR study in China aimed at 149 resistance genes, found that 63 were significantly increased in soils receiving swine manure containing antibiotics, with an average 192-fold enrichment (but as high as 28,000-fold); transposases were also increased (189-fold average enrichment, and as high as 90,000-fold) (Zhu et al., 2013). In another study, comparing with an unmanured soil, soils receiving swine manure had significantly more *erm*(A) (from 0% of the samples of unmanured, to 81% of the samples of manure tested positive) and *erm*(F) (form 6% to 62%) genes, but no difference was found in *erm*(B) genes (75% to 87%) (Marti et al., 2013b). The persistence of resistance genes applied along with manure can be of several months, but it seems to depend on climate conditions (Marti et al., 2014); this provides a (false) sense of safety that only considers the risk of resistant pathogens rapidly reaching foodstuff, but does not takes into account the vast, long-term disturbance of soil microbiota. In an interesting twist of this story, a report on the effect of applying cow manure (from non-medicated animals), was an increased presence of cephalothin resistant isolates; such isolates were not originated from the manure, but actually were an enriched fraction of resistant soil bacteria (main beta-lactamase, bla_{CEP-04} was found in the soil *before* manure treatment, and then enriched). Further analysis showed a significant shift in the structure of the affected soil microbiome, with *Pseudomonas* and *Janthinobacterium* spp. becoming more abundant due to the application of manure (Udikovik-Kolic et al., 2014).

Although factory farming is often blamed as the main source of resistant bacteria, as compared to grassland-based production systems, the latter somehow manage to also deliver resistance genes into the environment. While these small-scale farming activities use much less antibiotics, a study on Colombian farms showed that the weekly use of 13–130 g for 60–100 bovine animals, still caused a high prevalence of *tet*B(P), *tet*(Q), *tet*(W) and *tet*(O) genes in fecal samples (73–100% of samples positive), which end up in runoff water, ground water and soil around the farming areas. While *tet*B(P) genes mostly disappear in the environment, *tet*(Q) and *tet*(W) genes persist, being detectable in 43–67% of the soil and water samples (Santamaría et al., 2011).

Two further factors confound the analysis of the release of antibiotic resistant bacteria to the soil: (a) metagenomic approaches may not be very useful to understand the impact of human activities upon resistance in the soil: the overabundance of a variety of ancient genes already present (discussed in Chapter 2) may contaminate the results; a comparison between soils receiving manure or not indicated that about 70% of the detected resistance genes were related to manure application (Su et al., 2014); and (b) the presence of antibiotics in manure makes it difficult to discriminate between resistant bacteria within manure surviving in the soil; or soil bacteria that are being selected by the antibiotics in the manure, perhaps after receiving resistance genes via HGT.

4.2 RESISTANT BACTERIA IN ANIMALS

Farm or pet animals receiving antibiotics, therapeutically, prophylactically, or for growth promotion, have their microbiotas profoundly altered and, most significantly, enriched with resistant varieties of microbes. These direct effects, along with those caused by the clinical use of antibiotics upon the microbiotas of humans, would not be analyzed here, as they are rather obvious and the subject of clinical or veterinary microbiology surveys. However, the presence of resistant bacteria in wild animals not living in captivity demands a completely different explanation. Depending on the approach, finding antibiotic resistant bacteria, by culture-based methods; or antibiotic resistance genes, by molecular-based techniques, can be the result of a diverse array of conditions. Culturable, commensal bacteria from birds and mammals (the most frequently sampled kinds of wild animals) are clearly different from soil bacteria, hence are supposedly not under the influence of the purported presence of natural antibiotics in the soil. To find resistant varieties within this fraction of wildlife microbiota may indicate

(a) the acquisition of such resistant organism (or resistance gene) from a human-related source; (b) the selective effect of human-made antibiotics released into the environment; or (c) the presence of an unknown selective pressure, co-selecting for antibiotic resistance, in said environment. On the other hand, the detection of resistance genes, especially those not previously characterized in clinical isolates, may simply indicate a transient presence of soil bacteria (carrying the soil "resistome") within the sampled animal; it is also possible that, as with the metagenomic study of the microbiota of isolated human populations (see Chapter 1), there are plenty of "resistance" genes in unculturable bacteria from wildlife which, aside from being material for papers, have no further relevance whatsoever.

The detection of resistant *E. coli* in wildlife is not new: there are reports as old as 1968 for isolates from animals not known to be under antibiotic influence, and many more reports from diverse wildlife, starting from the late 1970's (Guenther et al., 2011). *E. coli* producing diverse ESBLs were first reported in 2006 (birds, deer, fox; 16.1% prevalence); since then all sorts of wild animals (gulls, boars, rats, owls, vultures, eagles, fish, etc.) have been found to carry ESBL-producing (mostly CTX-M, but also TEM and SHV) *E. coli* at rates ranging from 0.5 (German rats, 2009) to 32% (Portuguese seagulls, 2010) (Guenther et al., 2011). A study of resistance on *E. coli* isolates from feces of a variety of wild animals (elephants, warthogs, impalas and baboons comprising 70% of the samples) living freely but in proximity to humans in the Chobe District of Botswana, tested for resistance to ampicillin, ceftiofur, chloramphenicol, ciprofloxacin, doxycycline, gentamicin, neomycin, streptomycin, tetracycline and sulfamethoxazole-trimethoprim, showed that 41% of the isolates were resistant to at least one, and 35% to at least 3 antibiotics. Antibiotics with resistance prevalences above 10% were ampicillin, tetracycline, doxycycline, and streptomycin. While at much lower prevalence than in clinical isolates from the same region, resistance in animals' microbiota follow the same profile. Further analysis showed that carnivore animals and those living in or near water (crocodiles, hippopotamus, otters and waterbucks) had 2-to-6-fold higher rates of resistant and multi-resistant isolates; and those living within the protected Chobe National Park harbor much less multi-resistant bacteria than those living along urban or periurban human settlements (9 *vs.* 26%) (Jobbins and Alexander, 2015).

Birds have been the focus of great interest (particularly since Hitchcock, 1963), regarding their potential role in the spread of resistant bacteria. Being airborne, birds can cover wide areas, get frequently in and out human settlements and other contaminated areas, such as landfills, land and aquatic farms, wastewater treatment plants and manured soils; being omnivorous as a group, anything from antibiotic-containing animal food, to discarded antibiotics in landfills, to antibiotic-resistant bacteria in carcasses, can be taken up by birds, and then spread over large areas both urban and rural. The first report of ESBL-producing *E. coli* in wild birds was published in 2006, although other resistance phenotypes were reported in isolates from birds in 1975. Since then, reports of resistant bacteria from all kids of birds (ducks, geese, cormorants, gulls, doves, passerines, etc.), in most of the world (*e.g.*, Alaskan gulls carry plenty of *E. coli* and *K. pneumoniae* producers of CTX-M, SHV and TEM ESBLs (Bonnedahl et al., 2014)), and linked to human activity, bird migration, and many other conditions, have been reported (Bonnedahl and Järhult, 2014). A particularly robust, multi-national study of resistance prevalence in *E. coli* isolated from European gulls, showed a gradient of resistance levels corresponding to anthropogenic influence, either as human settlements (within Latvia, much higher resistance was found in samples from the capital city, Riga, than on the countryside), clinical antibiotic usage (the highest resistance rates were found in Spain), or use in food production animals (higher in The Netherlands than in Sweden). Isolates without detectable resistance ranged between 32.9% (Mazzaron, Spain) and 96.7% (Kaltene, Latvia); while isolates resistant to 7 drugs were mostly found, at 0.8–1.5% rates, in Spain and England. Ampicillin (up to 44.9%) and tetracycline (up to 52%) resistance were the most common phenotypes found (Stedt et al., 2014). Other interesting findings in birds include:

- A variety of *qnr* plasmid-mediated quinolone resistance genes in Enterobacteriaceae from US crows (being most common *qnrB47*, 28%; *qnrB6*, 27%; and *qnrB10*, 23%) (Halová et al., 2014).
- A 15.7% prevalence of cefotaxime-resistant *E. coli* in 65 sampled birds in The Netherlands, all carrying either an ESBL (mostly CTX-M) or an AmpC (*bla*$_{CMY-2}$, mostly plasmid-encoded); these isolates were also commonly resistant to sulfamethoxazole (57%), tetracycline (61%) and ciprofloxacin (48%) among other antibiotics, and 11% carried a plasmid-mediated quinolone resistance gene (Veldman et al., 2013).
- Birds of prey from Germany and Mongolia, two regions with very different population characteristics and antibiotic usage patterns, had essentially the same carriage of ESBL-producers, and the same rate of ESBL-producing *E. coli* (14% in Germany, 11% in Mongolia), differing only in the type of ESBL (CTX-M-1, 100% in Germany; CTX-M-9, 75% in Mongolia) (Guenther et al., 2012).
- ESBL-producing *E. coli* were found in 1.6% of 499 fecal or cloacal samples from great cormorants from Central Europe, and plasmid-mediated quinolone resistance genes in 1.2% of the same birds and 6% of mallards' samples; both kinds of genes were carried in conjugative plasmids (Tausova et al., 2012).

Many studies of resistance in wildlife microbiota focus on *E. coli* and/or ESBLs, as standard indicators of resistance. An interesting exception is an analysis of the fecal microbiota of howler and spider monkeys, tapirs and felids (jaguars, ocelots) from Mexican forests (Cristóbal-Azkarate et al., 2014). This culture-based study included many different bacterial species, and several antibiotics; it was also capable of establishing the relationship between resistance prevalence and a geographic gradient of human influence. Fresh feces were sampled, and precise coordinates were registered; samples were briefly pre-cultured in liquid media (allowing for recovery of transport stress), plated on selective plates (containing antibiotics at resistance breakpoint concentrations, according to CLSI), as well as non-selective and indicative plates. Bacteria growing on selective plates were isolated, biochemically identified, and their susceptibility to a panel of antibiotics carried out by disk diffusion. Additional testing included a phenotypic assay for distinguishing aminoglycoside-resistance mechanisms, and PCR-based detection of class-1 integrons. Resistance phenotypes deemed as "intrinsic" (*e.g.*, ampicillin-resistance in *Pseudomonas* and *Acinetobacter*, cephalosporin-resistance in enterococci) were excluded from the analysis. Many Enterobacteriaceae (*E. coli, Klebsiella, Enterobacter, Citrobacter*) and non-fermentative bacteria (*Pseudomonas, Acinetobacter*) were isolated, and a few gram-positives. Resistance to ampicillin, tetracycline, chloramphenicol and sulfonamides were the most frequently found. In order to simplify the analysis, some indicators were used for each type of sampled animal and distance from human settlements: proportion of samples with at least one resistant isolate (rS), number of isolates per sample that were resistant to at least one drug (rO), total number of resistance phenotypes per sample (rP), and average number of resistance phenotypes per isolate (rA). Results were somewhat contradictory and highlighted the importance of looking beyond *E. coli*. Among the expected results, terrestrial animals (tapirs and felids) had more resistance than arboreal ones (monkeys), measured as rS and rO, suggesting a terrestrial route for resistance, likely water-related; samples collected at ≤2.5 km from human settlements had higher rO and rP; and samples from howler monkeys living in an anthropogenically more disturbed region had higher rS, rO and rP than samples from the same species living in a more conserved region, and enzyme-mediated resistance to gentamicin (likely to be plasmid-borne), plasmid-mediated beta-lactamases, and presence of class-1 integrons, were also found more often in samples from the disturbed region. But when looking only at *E. coli*, results were rather the opposite: much less resistance was found in isolates from howler monkeys living in disturbed areas than in those from conserved ones; and isolates from terrestrial animals were less resistant than those from arboreal (except for tetracycline resistance in tapirs). Furthermore, resistance to ciprofloxacin was only found in samples from regions without significant human influence, and ESBL-mediated resistance to cefotaxime was found all across the sampled region, although at low rate (10 isolates out of 247) and only at arboreal, non-*E. coli* isolates. A "resistance index" for each sample, not included in the paper (some co-authors were afraid of the prudish peer-review), with a formula $\sum rP \times \sum rQ \times \sum rO$, where rQ was a subjective qualification of the relevance of each resistance phenotype (1 for ampicillin and tetracycline, 2 for amoxicillin-clavulanate, chloramphenicol and sulfonamides, and 3 for cefotaxime, gentamicin and ciprofloxacin), yielded additional insights: again, samples from terrestrial animals (average 174, 119 for tapirs, 283 for felids; interestingly, the top-of-the-food-chain seem to concentrate resistant organisms as in the Botswana study above) were higher than those from arboreal ones (average 74, 79 for howler monkeys and 69 for spider monkeys); and without significant difference between howler monkeys from conserved (79) or disturbed (76) regions. This index could be a useful tool when analyzing complex data that include isolates of different species, from different animals, from different regions.

Other studies looking beyond *E. coli* are scarce:

– A single *Enterococcus faecium* carrying a *vanA* gene, along with *erm*(A), *erm*(B), *tet*(M), *dfrG* and *dfrK* genes, was detected in a Spanish wild boar, among 348 cloacal or rectal samples from wild birds and boars; all other vancomycin-resistant enterococcal isolates were considered to be intrinsical phenotypes (Lozano et al., 2015).

– Enterococci from Chilean camelids were also mostly susceptible to penicillin, daptomycin, levofloxacin, erythromycin, linezolid, minocycline and nitrofurantoin, and glycopeptide resistance was found in 3 *E. casseliflavus* isolates (Guerrero-Olmos et al., 2014).

- Migratory rooks sampled in Austria carried MRSA (5 out of 54 samples; two with SCCmec IVa and three with SCCmec IVc; (Loncaric et al., 2013b)), as did European otters and brown hares (Loncaric et al., 2013a).
- *Salmonella* carriage in wild boars is increased if the animals co-habited with cattle, but the two only resistant isolates (a *S.* Mbandaka, resistant to sulfamethoxazole, streptomycin and chloramphenicol; and a *S.* Enteritidis, resistant to ciprofloxacin and, obviously, to nalidixic acid) were isolated from cattle-linked and cattle-free areas, one each (Navarro-Gonzalez et al., 2012).

Animals become unsuspected pathways for resistant bacteria into the environment: for instance, wild animals being captured by illegal traffickers, when rescued and reintroduced into their natural habitats, carry with them bacteria acquired during their cruel and brief contact with humans (Braconaro et al., 2015). And insects, particularly flies and cockroaches, that carry a wide variety of bacteria, including important human pathogens, have also been reported as carriers for multi-resistant *E. coli* and enterococci, mostly from swine and poultry farms and wastewater treatment plants (Zurek and Ghosh, 2014).

4.3 ORGANISMS, GENES, ROUTES: MORE THAN "JUST" RESISTANCE

While this book is mainly focused on antibiotics and antibiotic resistance, it is clear that they cannot be analyzed separately from other agents affecting microbial communities; and from other particular features of bacterial biology. As was reviewed in the previous chapter, in addition to antibiotics, several related, human-made selective pressures, are released into the environment in copious amounts; disinfectants and heavy metals have been studied in detail, but other agents may have relevant impact as modifiers of bacterial susceptibility to antibiotics. Non-antibiotic drugs, that are also present in wastewaters, are known to have limited, but detectable antimicrobial properties, that could potentially select for unspecific resistance phenotypes. Furthermore, the impact of antibiotics themselves is far from only selecting antibiotic resistance, which is a rather obvious inference from both, the low concentrations of antibiotics released by many known sources, and the biological effects of sub-inhibitory concentrations of antibiotics. When growing biofilms from river waters, and then exposing them to antibiotics (erythromycin, sulfonamides) or gemfibrozil (a hypolipidemic agent), at concentrations commonly found in rivers as a consequence of wastewater contamination, different changes occurred to them: erythromycin induced a gene involved in the synthesis of guanosine tetraphosphate (ppGpp) and of heat shock protein DnaK; sulfonamides induced the expression of DNA- and RNA-polymerases, while other responses varied depending on the sulfonamide used; and gemfibrozil affected genes related to lipid metabolism and of flagellar synthesis, which in turn could affect biofilm formation (Yergeau et al., 2010). These results highlight the various effects that even minute concentrations of antibiotics and other drugs can exert upon natural microbial populations in open environments. The very composition of biofilms in water bodies is affected by the input of bacteria from wastewater treatment plants. While most bacteria in biofilms from a river upstream of a treatment plant were Firmicutes, the composition of biofilms downstream included more Alpha-Beta- and particularly more Gamma-proteobacteria, a composition akin to the one of treatment plant biofilms (Marti et al., 2013a). Regardless of the presence of resistance genes, the massive release of wastewater bacteria into the environment is dramatically changing microbial ecology, with consequences that are very hard to predict. As to disinfectants, triclosan promotes the binding of *S. aureus* to inanimate surfaces (glass, plastic; as well as to human proteins, promoting nasal colonization (Syed et al., 2014)); such an effect, additional to its biocidal properties, could also modify the composition of microbial communities in environments receiving wastewater containing triclosan.

Antibiotic metabolites and decay products deserve a separate paragraph. When assessing the environmental concentrations of antibiotics, analysis seldom include antibiotic metabolites, hence their effects are neglected; furthermore, as many metabolites have diminished or null antibacterial properties, they are not considered as a relevant influence upon microbial communities. However, this notion is not entirely true: anhydroerythromycin, a metabolite of erythromycin that is excreted in the urine, mostly lacks antimicrobial activity; however, it is still capable of inducing *erm* genes and, consequently, of conferring an MLS_B phenotype in staphylococci, at concentrations present in wastewater (1 ng/L) (Heß and Gallert, 2014). Likewise, tetracycline decay products (epitetracycline, epianhydrotetracycline) are present in the environment, and persist much longer than the precursor molecule. Decay products strongly induce the expression of *tet* genes (Palmer, 2012). This is actually a desirable effect: by unnecessarily inducing the expression of resistance mechanisms, these by-products select *against* resistance, with resistance becoming a disadvantage in the absence of inhibitory concentrations of antibiotics. Much more research is needed to fully understand this apparent paradox.

Other chemicals that are used in huge amounts for agricultural purposes, but that have seldom been analyzed as contributors to the bacterial resistance issue, are herbicides. For instance, with ~8,300 tons/year, glyphosate is the most used agricultural herbicide in the US (plus 500 tons/year for home gardens); the use of these agents is likely to increase significantly, with upcoming genetically modified crops whose "advantages" depend on the use of these chemicals. In a recent study, concentrations of glyphosate, dicamba and 2,4-dichlorophenoxyacetic acid (2,4-D) below those of recommended use, modified the susceptibility of *E. coli* and *S.* Typhimurium to a number of antibiotics, increasing it to some but decreasing it to most. The activity of each herbicide upon each species varied, but in general dicamba diminished susceptibility to chloramphenicol, ciprofloxacin and tetracycline, while increasing it to kanamycin; 2,4-D diminished susceptibility to ampicillin, ciprofloxacin, chloramphenicol and tetracycline, while increasing it to kanamycin; and glyphosate diminished susceptibility to ciprofloxacin and kanamycin, while increasing it to chloramphenicol and tetracycline. Some of these transient changes, such as decreased susceptibility of *E. coli* to chloramphenicol induced by dicamba, and to kanamycin induced by glyphosate, were reversed by an efflux-pump inhibitor; and dicamba treatment induced the expression of the *soxRS* regulon (Kurenbach et al., 2015). In fact, one of the most commonly used herbicides worldwide, paraquat (it is difficult to know how much paraquat is actually used but sales amount up to US$396–430 millions yearly), is a known inducer of the *soxRS* regulon of *E. coli*, whose activation transiently reduces the susceptibility to a number of antibiotics, mostly due to decreased permeability and increased efflux (Amábile-Cuevas and Demple, 1991). While all these effects fall within the "adaptive resistance" category, it is likely that the activation of these responses contribute to the acquisition of full-resistance under some circumstances, that will be discussed in the following chapter.

From the biological side of the equation, resistance genes in the environment are thought to be mostly within bacterial cells, regardless of their ability to express as a resistance phenotype. However, they can be present within phages, which dramatically increases their potential for horizontal spread. A Spanish study of phage DNA obtained from city (Barcelona) sewage and a river (Llobregat) receiving human and animal waste, found an average of 1,000–10,000 copies of beta-lactamase *bla*TEM gene per mL, ~10 copies of *bla*CTX-M per mL, and nearly 100 copies of *mecA* (conferring methicillin resistance in *S. aureus*) per mL. Gene counts were 1–2 orders of magnitude lower in phage DNA than in bacterial DNA (Colomer-Lluch et al., 2011b). The same genes were detected in phage DNA from fecal material, slurry and wastewater of cattle, pig and poultry farms, in amounts of 100–1,000 gene copies per mL or g, for *bla*CTX-M and *mecA*, but of 1,000–100,000 gene copies per mL or g for *bla*TEM (Colomer-Lluch et al., 2011a).

In addition to resistant bacteria and resistance genes, human influence seems to increase the overall presence of mobile genetic elements in bacteria:

- *E. coli* isolates from human and animal hosts in cities carried significantly more plasmids that those from wildlife (Souza et al., 1999).
- *Vibrio* spp. isolated from a polluted river in China carry diverse ICEs encoding resistance and virulence genes not reported before (Song et al., 2013).
- Class 1 integrons are present at higher amounts in environmental samples from "anthropogenically impacted" zones: *e.g.*, from 0–6% upstream of wastewater treatment plants, to 8–86% downstream; 3.6–30% in lake estuaries, 89.3% in manured soils; and in numbers that are often in the 10^2–10^4 copies/L, but as high as 10^{11}/g in lake sediments (Stalder et al., 2012).
- In samples downstream of a wastewater treatment plant processing effluents from drug manufacturing factories in India, having ciprofloxacin concentrations of 0.1–1 mg/g of organic matter in the sediments, *intI1* genes accounted for ~0.2% of identified bacterial genes, and a class 2 transposase associated with insertion sequence common regions for 0.4–0.8% (Kristiansson et al., 2011).
- A variety of integrase genes, including 14 previously unknown ones, were detected using degenerate priming PCR on heavy-metal contaminated mine tailings (Nemergut et al., 2004).

By increasing the presence of mobile genetic elements, human influence, antibiotic-derived or not, may be fostering the ability of bacteria to exchange genes and even to put in motion that ancient "resistome" that, so far, has been mostly circumscribed to soil bacteria.

Heavy metals are known to exert toxic effects upon bacteria, and are also released in huge amounts in the environment (see previous chapter). This is relevant to the issue at hand because:

- There are genes conferring cross-resistance to antibiotics and heavy metals, mostly encoding efflux pumps.
- Many heavy-metal resistance genes are linked to antibiotic resistance genes, such as the *tcrB* gene conferring Cu resistance, found along *erm*(B) and *vanA* in *E. faecium*; *mer* operons conferring resistance to Hg as a component of Tn*21* transposons also carrying class-1 integrons; and *czcA*, encoding an efflux pump known to extrude Zn, Cd and Co, is also located closely to class-1 integrons in fresh water bacteria, to cite a few.
- Cr and Cu (and Hg (Fuentes and Amábile-Cuevas, 1997)) can induce the expression of the *soxRS* regulon of *E. coli*, that includes the AcrAB efflux pump conferring resistance to a number of antibiotics (Seiler and Berendonk, 2012).

Detection of heavy metal resistance genes in the environment has not been explored nearly as intensively as antibiotic resistance ones; the use of metal-resistant organisms for bioremediation (Haferburg and Kothe, 2007) is particularly worrisome, if they are to be released into the environment, potentially fostering the gene pool allowing for co-selection of antibiotic resistance.

As has been reviewed up to this chapter, there is a number of actual and putative resistance genes in soil bacteria that have never been detected in clinically-relevant bacteria; antibiotics and a number of known and potential selective pressures are discharged, in enormous quantities, into waters and soils, along with commensal and pathogenic bacteria carrying antibiotic resistance genes and mobile genetic elements, along with other determinants that may enable co-selection and other evolutive processes. In the following chapter, the known and hypothetical consequences of this dreadful cocktail will be discussed.

REFERENCES

Agersø, Y. & Sandvang, D. (2005) Class 1 integrons and tetracycline resistance genes in alcaligenes, arthrobacter, and *Pseudomonas* spp. isolated from pigsties and manured soil. *Appl. Environ. Microbiol.*, 71, 7941–7947.

Akoachere, J. F. T. K., Masalla, T. N. & Njom, H. A. (2013) Multi-drug resistant toxigenic *Vibrio cholerae* O1 is persistent in water sources in New Bell-Couala, Cameroon. *BMC Infect. Dis.*, 13, 366.

Amábile-Cuevas, C. F. (2010) Antibiotic resistance in Mexico: a brief overview of the current status and its causes. *J. Infect. Dev. Ctries.*, 4, 126–131.

Amábile-Cuevas, C. F., Arredondo-García, J. L., Cruz, A. & Rosas, I. (2009) Fluoroquinolone resistance in clinical and environmental isolates of *Escherichia coli* in Mexico City. *J. Appl. Microbiol.*, 108, 158–162.

Amábile-Cuevas, C. F. & Demple, B. (1991) Molecular characterization of the *soxRS* genes of *Escherichia coli*: two genes control a superoxide stress regulon. *Nucleic Acids Res.*, 19, 4479–4484.

Amos, G. C. A., Gozzard, E., Carter, C. E., Mead, A., Bowes, M. J., Hawkey, P. M., Zhang, L., Singer, A. C., Gaze, W. H. & Wellington, E. M. H. (2015) Validated predictive modelling of the environmental resistome. *ISME J.*, 10.1038/ismej.2014.237.

Amos, G. C. A., Zhang, L., Hawkey, P. M., Gaze, W. H. & Wellington, E. M. (2014) Functional metagenomic analysis reveals rivers are a reservoir for diverse antibiotic resistance genes. *Vet. Microbiol.*, 171, 441–447.

Ash, R. J., Mauck, B. & Morgan, M. (2002) Antibiotic resistance of gram-negative bacteria in rivers, United States. *Emerg. Infect. Dis.*, 8, 713–716.

Aubron, C., Poirel, L., Ash, R. J. & Nordmann, P. (2005) Carbapenemase-producing Enterobacteriaceae, U.S. rivers. *Emerg. Infect. Dis.*, 11, 260–264.

Bhullar, K., Waglechner, N., Pawlowski, A., Koteva, K., Banks, E. D., Johnston, M. D., Barton, H. A. & Wright, G. D. (2012) Antibiotic resistance is prevalent in an isolated cave microbiome. *PLoS One*, 7, e34953.

Bonnedahl, J., Hernandez, J., Stedt, J., Waldenström, J., Olsen, B. & Drobni, M. (2014) Extended-spectrum β-lactamases in *Escherichia coli* and *Klebsiella pneumoniae* in gulls, Alaska, USA. *Emerg. Infect. Dis.*, 20, 897–899.

Bonnedahl, J. & Järhult, J. D. (2014) Antibiotic resistance in wild birds. *Upsala J. Med. Sci.*, 119, 113–116.

Braconaro, P., Saidenberg, A. B., Benites, N. R., Zuniga, E., Da Silva, A. M., Sanches, T. C., Zwarg, T., Brandão, P. E. & Melville, P. A. (2015) Detection of bacteria and fungi and assessment of the molecular aspects and resistance of *Escherichia coli* isolated from confiscated passerines intended for reintroduction programs. *Microb. Pathog.*, in press.

Broaders, E., Gahan, C. G. M. & Marchesi, J. R. (2013) Mobile genetic elements of the human gastrointestinal tract. *Gut Microbes*, 4, 271–280.

Casellas, J. M. & Quinteros, M. G. (2007) A Latin American "point de vue" on the epidemiology, control, and treatment options of infections caused by extended-spectrum beta-lactamase producers. In Amábile-Cuevas, C. F. (Ed.) *Antimicrobial resistance in bacteria.* Wymondham, Horizon Bioscience.

Chakraborty, R., Kumar, A., Bhowal, S. S., Mandal, A. K., Tiwary, B. K. & Mukherjee, S. (2013) Diverse gene cassettes in class 1 integrons of facultative oligotrophic bacteria of river Mahananda, West Bengal, India. *PLoS One*, 8, e71753.

Chen, B., Yang, Y., Liang, X., Yu, K., Zhang, T. & Li, X. (2013) Metagenomic profiles of antibiotic resistance genes (ARGs) between human impacted estuary and deep ocean sediments. *Environ. Sci. Technol.*, 47, 12753–12760.

Collignon, P., Athukorala, P., Senanayake, S. & Khan, F. (2015) Antimicrobial resistance: the major contribution of poor governance and corruption to this growing problem. *PLoS One*, 10, e0116746.

Colomer-Lluch, M., Imamovic, L., Jofre, J. & Muniesa, M. (2011a) Bacteriophages carrying antibiotic resistance genes in fecal waste from cattle, pigs, and poultry. *Antimicrob. Agents Chemother.*, 55, 4908–4911.

Colomer-Lluch, M., Jofre, J. & Muniesa, M. (2011b) Antibiotic resistance genes in the bacteriophage DNA fraction of environmental samples. *PLoS One*, 6, e17549.

Cotton, M. F., Wasserman, E., Pieper, C. H., Theron, D. C., Van Tubbergh, D., Campbell, G., Fang, F. C. & Barnes, J. (2000) Invasive disease due to extended spectrum beta-lactamase-producing *Klebsiella pneumoniae* in a neonatal unit: the possible role of cockroaches. *J. Hosp. Infect.*, 44, 13–17.

Cristóbal-Azkarate, J., Dunn, J. C., Day, J. M. W. & Amábile-Cuevas, C. F. (2014) Resistance to antibiotics of clinical relevance in the fecal microbiota of Mexican wildlife. *PLoS One*, 9, e107719.

Czekalski, N., Berthold, T., Caucci, S., Egli, A. & Bürgmann, H. (2012) Increased levels of multiresistant bacteria and resistance genes after wastewater treatment and their dissemination into Lake Geneva, Switzerland. *Front. Microbiol.*, 3, 106.

D'Costa, V. M., King, C. E., Kalan, L., Morar, M., Sung, W. W. L., Schwarz, C., Froese, D., Zazula, G., Calmels, F., Debruyne, R., Golding, G. B., Poinar, H. N. & Wright, G. D. (2011) Antibiotic resistance is ancient. *Nature*, 477, 457–461.

Dhanji, H., Murphy, N. M., Akhigbe, C., Doumith, M., Hope, R., Livermore, D. M. & Woodford, N. (2011) Isolation of fluoroquinolone-resistant O25b:H4-ST131 *Escherichia coli* with CTX-M-14 extended-spectrum β-lactamase from UK river water. *J. Antimicrob. Chemother.*, 66, 512–516.

Díaz-Mejía, J. J., Amábile-Cuevas, C. F., Rosas, I. & Souza, V. (2008) An analysis of the evolutionary relationships of integron integrases, with emphasis on the prevalence of class 1 integron in *Escherichia coli* isolates from clinical and environmental origins. *Microbiology*, 154, 94–102.

Forslund, K., Sunagawa, S., Kultima, J. R., Mende, D. R., Arumugam, M., Typas, A. & Bork, P. (2013) Country-specific antibiotic use practices impact the human gut resistome. *Genome Res.*, 23, 1163–1169.

Foxman, B. (2003) Epidemiology of urinary tract infections: incidence, morbidity, and economic costs. *Dis. Mon.*, 49, 53–70.

Fuentes, A. M. & Amábile-Cuevas, C. F. (1997) Mercury induces multiple antibiotic resistance in *Escherichia coli* through activation of SoxR, a redox-sensing regulatory protein. *FEMS Microbiol. Lett.*, 154, 385–388.

Gao, L., Hu, J., Zhang, X., Wei, L., Li, S., Miao, Z. & Chai, T. (2015) Application of swine manure on agricultural fields contributes to extended-spectrum β-lactamase-producing *Escherichia coli* spread in Tai'an, China. *Front. Microbiol.*, 14, 313.

Gatica, J. & Cytryn, E. (2013) Impact of treated wastewater irrigation on antibiotic resistance in the soil microbiome. *Environ. Sci. Pollut. Res. Int.*, 20, 3529–3538.

Ghosh, A., Kukanich, K., Brown, C. E. & Zurek, L. (2012) Resident cats in small animal veterinary hospitals carry multi-drug resistant enterococci and are likely involved in cross-contamination of the hospital environment. *Front. Microbiol.*, 3, 62.

Guardabassi, L., Petersen, A., Olsen, B. & Dalsgaard, A. (1998) Antibiotic resistance in *Acinetobacter* spp. isolated from sewers receiving waste effluent from a hospital and a pharmaceutical plant. *Appl. Environ. Microbiol.*, 64, 3499–3502.

Guenther, S., Aschenbrenner, K., Stamm, I., Bethe, A., Semmler, T., Stubbe, A., Stubbe, M., Batsajkhan, N., Glupczynski, Y., Wieler, L. H. & Ewers, C. (2012) Comparable high rates of extended-spectrum-beta-lactamase-producing *Escherichia coli* in birds of prey from Germany and Mongolia. *PLoS One*, 7, e53039.

Guenther, S., Ewer, C. & Wieler, L. H. (2011) Extended-spectrum beta-lactamases producing *E. coli* in wildlife, yet another form of environmental pollution? *Front. Microbiol.*, 2, 246.

Guerrero-Olmos, K., Báez, J., Valenzuela, N., Gahona, J., Del Campo, R. & Silva, J. (2014) Molecular characterization and antibiotic resistance of *Enterococcus* species from gut microbiota of Chilean Altiplano camelids. *Infect. Ecol. Epidemiol.*, 4, 24714.

Haferburg, G. & Kothe, E. (2007) Microbes and metals: interactions in the environment. *J. Basic Microbiol.*, 47, 453–467.

Halová, D., Papousek, I., Jamborova, I., Masarikova, M., Cizek, A., Janecko, N., Oravcova, V., Zurek, L., Clark, A. B., Townsend, A., Ellis, J. C. & Literak, I. (2014) Plasmid-mediated quinolone resistance genes in Enterobacteriaceae from American crows: high prevalence of bacteria with variable *qnrB* genes. *Antimicrob. Agents Chemother.*, 58, 1257–1258.

Hes, S. & Gallert, C. (2014) Demonstration of staphylococci with inducible macrolide-lincosamide-streptogramin B (MLS$_B$) resistance in sewage and river water and of the capacity of anhydroerythromycin to induce MLS$_B$. *FEMS Microbiol. Ecol.*, 88, 48–59.

Heuer, H., Solehati, Q., Zimmerling, U., Kleineidam, K., Schloter, M., Müller, T., Focks, A., Thiele-Bruhn, S. & Smalla, K. (2011) Accumulation of sulfonamide resistance genes in arable soils due to repeated application of manure containing sulfadiazine. *Appl. Environ. Microbiol.*, 77, 2527–2530.

Jardine, C. M., Janecko, N., Allan, M., Boerlin, P., Chalmers, G., Kozak, G., Mcewen, S. A. & Reid-Smith, R. J. (2012) Antimicrobial resistance in *Escherichia coli* isolates from raccoons (*Proyon lotor*) in Southern Ontario, Canada. *Appl. Environ. Microbiol.*, 78, 3873–3879.

Jechalke, S., Broszat, M., Lang, F., Siebe, C., Smalla, K. & Grohmann, E. (2015) Effects of 100 years wastewater irrigation on resistance genes, class 1 integrons and IncP-1 plasmids in Mexican soil. *Front. Microbiol.*, 6, 163.

Jobbins, S. E. & Alexander, K. A. (2015) Whence they came -antibiotic-resistant *Escherichia coli* in African wildlife. *J. Wildl. Dis.*, 51, 1–10.

Kelch, W. J. & Lee, J. S. (1978) Antibiotic resistance patterns of gram-negative bacteria isolated from environmental sources. *Appl. Environ. Microbiol.*, 36, 450–456.

King, G. M. (2014) Urban microbiomes and urban ecology: how do microbes in the built environment affect human sustainability in cities? *J. Microbiol.*, 52, 721–728.

Kristiansson, E., Fick, J., Janzon, A., Grabic, R., Rutgersson, C., Weijdegard, B., Söderström, H. & Larsson, D. G. J. (2011) Pyrosequencing of antibiotic-contaminated river sediments reveals high levels of resistance and gene transfer elements. *PLoS One*, 6, e17038.

Kurenbach, B., Marjoshi, D., Amábile-Cuevas, C. F., Ferguson, G. C., Godsoe, W., Gibson, P. & Heinemann, J. A. (2015) Sublethal exposure to commercial formulations of the herbicides dicamba, 2,4-dichlorophenoxyacetic acid, and glyphosate cause changes in antibiotic susceptibility in *Escherichia coli* and *Salmonella enterica* serovar Typhimurium. *mBio*, 6, e00009-15.

Kyselková, M., Jirout, J., Vrchotová, N., Schmitt, H. & Elhottová, D. (2015) Spread of tetracycline resistance genes at a conventional dairy farm. *Front. Microbiol.*, 29, 536.

Lapara, T. & Burch, T. (2012) Municipal wastewater as a reservoir of antibiotic resistance. In Keen, P. L. & Montforts, M. H. M. M. (Eds.) *Antimicrobial resistance in the environment*. Hoboken, John Wiley & Sons.

Laroche, E., Pawlak, B., Berthe, T., Skurnik, D. & Petit, F. (2009) Occurrence of antibiotic resistance and class 1, 2 and 3 integrons in *Escherichia coli* isolated from a densely populated estuary (Seine, France). *FEMS Microbiol. Ecol.*, 68, 118–130.

Leclercq, R., Oberlé, K., Galopin, S., Cattoir, V., Budzinski, H. & Petit, F. (2013) Changes in enterococcal populations and related antibiotic resistance along a medical center – wastewater treatment plant – river contunuum. *Appl. Environ. Microbiol.*, 79, 2428–2434.

Li, D., Yu, T., Zhang, Y., Yang, M., Li, Z., Liu, M. & Qi, R. (2010) Antibiotic resistance characteristics of environmental bacteria from an oxytetracycline production wastewater treatment plant and the receiving river. *Appl. Environ. Microbiol.*, 76, 3444–3451.

Loncaric, I., Kübber-Heiss, A., Posautz, A., Stalder, G. L., Hoffmann, D., Rosengarten, R. & Walzer, C. (2013a) Characterization of methicillin-resistant *Staphylococcus* spp. carrying the *mecC* gene, isolated from wildlife. *J. Antimicrob. Chemother.*, 68, 2222–2225.

Loncaric, I., Stalder, G. L., Mehinagic, K., Rosengarten, R., Hoelzl, F., Knauer, F. & Walzer, C. (2013b) Comparison of ESBL – and AmpC producing Enterobacteriaceae and methicillin-resistant *Staphylococcus aureus* (MRSA) isolated from migratory and resident population of rooks (*Corvus frugilegus*) in Austria. *PLoS One*, 8, e84048.

Lozano, C., Gonzalez-Barrio, D., Camacho, M. C., Lima-Barbero, J. F., De La Puente, J., Höfle, U. & Torres, C. (2015) Characterization of fecal vancomycin-resistant enterococci with acquired and intrinsic resistance mechanisms in wild animals, Spain. *Microb. Ecol.*, in press.

Lu, S. Y., Zhang, Y. L., Geng, S. N., Li, T. Y., Ye, Z. M., Zhang, D. S., Zou, F. & Zhou, H. W. (2010) High diversity of extended-spectrum beta-lactamase-producing bacteria in an urban river sediment habitat. *Appl. Environ. Microbiol.*, 76, 5972–5976.

Maron, D. F., Smith, T. J. S. & Nachman, K. E. (2013) Restrictions on antimicrobial use in food animal production: an international regulatory and economic survey. *Globalization Health*, 9, 48.

Marti, E., Jofre, J. & Balcazar, J. L. (2013a) Prevalence of antibiotic resistance genes and bacterial community composition in a river influenced by a wastewater treatment plant. *PLoS One*, 8, e78906.

Marti, R., Scott, A., Tien, Y. C., Murray, R., Sabourin, L., Zhang, Y. & Topp, E. (2013b) Impact of manure fertilization on the abundance of antibiotic-resistant bacteria and frequency of detection of antibiotic resistance genes in soil and on vegetables at harvest. *Appl. Environ. Microbiol.*, 79, 5701–5709.

Marti, R., Tien, Y. C., Murray, R., Scott, A., Sabourin, L. & Topp, E. (2014) Safely coupling livestock and crop production systems: how rapidly do antibiotic resistance genes dissipate in soil following a commercial application of swine or daily manure? *Appl. Environ. Microbiol.*, 80, 3258–3265.

Murray, G. E., Tobin, R. S., Junkins, B. & Kushner, D. J. (1984) Effect of chlorination on antibiotic resistance profiles of sewage-related bacteria. *Appl. Environ. Microbiol.*, 48, 73–77.

Navarro-Gonzalez, N., Mentaberre, G., Porrero, C. M., Serrano, E., Mateos, A., López-Martín, J. M., Lavín, S. & Domínguez, L. (2012) Effect of cattle on *Salmonella* carriage, diversity and antimicrobial resistance in free-ranging wild boar (*Sus scrofa*) in northeastern Spain. *PLoS One*, 7, e51614.

Nemergut, D. R., Martin, A. P. & Schmidt, S. K. (2004) Integron diversity in heavy-metal-contaminated mine tailings and inferences about integron evolution. *Appl. Environ. Microbiol.*, 70, 1160–1168.

Novais, C., Coque, T. M., Ferreira, H., Sousa, J. C. & Peixe, L. (2005) Environmental contamination with vancomycin-resistant enterococci from hospital sewage in Portugal. *Appl. Environ. Microbiol.*, 71, 3364–3368.

Palmer, A. C. (2012) *Gene-drug interactions and the evolution of antibiotic resistance*, Cambridge, MA, Harvard University.

Ramírez Castillo, F. Y., Avelar González, F. J., Gameau, P., Márquez Díaz, F., Guerrero Barrera, A. L. & Harel, J. (2013) Presence of multi-drug resistant pathogenic *Escherichia coli* in the San Pedro River located in the State of Aguascalientes, Mexico. *Front. Microbiol.*, 4, 147.

Rosas, I., Amábile-Cuevas, C. F., Calva, E. & Osornio-Vargas, A. (2011) Animal and human waste as components of urban dust pollution: health implications. In Nriagu, J. O. (Ed.) *Encyclopedia of environmental health*. Amsterdam, Elsevier.

Rosas, I., Salinas, E., Martínez, L., Calva, E., Cravioto, A., Eslava, C. & Amábile-Cuevas, C. F. (2006) Urban dust fecal pollution in Mexico City: antibiotic resistance and virulence factors of *Escherichia coli*. *Int. J. Hyg. Environ. Health*, 209, 461–470.

Rosas, I., Salinas, E., Martínez, L., Cruz-córdova, A., González-Pedrajo, B., Espinosa, N. & Amábile-Cuevas, C. F. (2015) Characterization of *Escherichia coli* isolates from an urban lake receiving water from a wastewater treatment plant in Mexico City: fecal pollution and antibiotic resistance. *Curr. Microbiol.*, 71, 490–495.

Santamaría, J., López, L. & Soto, C. Y. (2011) Detection and diversity evaluation of tetracycline resistance genes in grassland-based production systems in Colombia, South America. *Front. Microbiol.*, 2, 252.

Schlüter, A., Szczepanowski, R., Pühler, A. & Top, E. M. (2007) Genomics of IncP-1 antibiotic resistance plasmids isolated from wastewater treatment plants provides evidence for a widely accessible drug resistance gene pool. *FEMS Microbiol. Rev.*, 31, 449–477.

Schwartz, T., Kohnen, W., Jansen, B. & Obst, U. (2003) Detection of antibiotic-resistant bacteria and their resistance genes in wastewater, surface water, and drinking water biofilms. *FEMS Microbiol. Ecol.*, 43, 325–335.

Seiler, C. & Berendonk, T. U. (2012) Heavy metal driven co-selection of antibiotic resistance in soil and water bodies impacted by agriculture and aquaculture. *Front. Microbiol.*, 3, 399.

Slekovec, C., Plantin, J., Cholley, P., Thouverez, M., Talon, D., Bertrand, X. & Hocquet, D. (2012) Tracking down antibiotic-resistant *Pseudomonas aeruginosa* isolates in a wastewater network. *PLoS One,* 7, e49300.

Song, Y., Yu, P., Li, B., Pan, Y., Zhang, X., Cong, J., Zhao, Y., Wang, H. & Chen, L. (2013) The mosaic accessory gene structures of the SXT/R391-like integrative and conjugative elements derived from *Vibrio* spp. isolated from aquatic products and environment in the Yangtze River estuary, China. *BMC Microbiol,* 13, 214.

Souza, V., Rocha, M., Valeria, A. & Eguiarte, L. E. (1999) Genetic structure of natural populations of *Escherichia coli* in wild hosts on different continents. *Appl. Environ. Microbiol.,* 65, 3373–3385.

Stalder, T., Barraud, O., Casellas, M., Dagot, C. & Ploy, M. C. (2012) Integron involvement in environmental spread of antibiotic resistance. *Front. Microbiol.,* 3, 119.

Stedt, J., Bonnedahl, J., Hernandez, J., Mcmahon, B. J., Hasan, B., Olsen, B., Drobni, M. & Waldenström, J. (2014) Antibiotic resistance patterns in *Escherichia coli* from gulls in nine European countries. *Infect. Ecol. Epidemiol.,* 4, 21565.

Stephen, A. M. & Cummings, J. H. (1980) The microbial contribution to human faecal mass. *J. Med. Microbiol.,* 13, 45–56.

Su, J. Q., Wei, B., Xu, C. Y., Qiao, M. & Zhu, Y. G. (2014) Functional metagenomic characterization of antibiotic resistance genes in agricultural soils from China. *Environ. Int.,* 65, 9–15.

Syed, A. K., Ghosh, S., Love, N. G. & Boles, B. R. (2014) Triclosan promotes *Staphylococcus aureus* nasal colonization. *mBio,* 5, e01015-13.

Szczepanowski, R., Braun, S., Riedel, V., Schneiker, S., Krahn, I., Pühler, A. & Schlüter, A. (2005) The 120592 bp IncF plasmid pRSB107 isolated from a sewage-treatment plant encodes nine different antibiotic-resistance determinants, two iron-acquisition systems and other putative virulence-associated functions. *Microbiology,* 151, 1095–1111.

Szczepanowski, R., Linke, B., Krahn, I., Gartemann, K. H., Gützkow, T., Eichler, W., Pühler, A. & Schlüter, A. (2009) Detection of 140 clinically relevant antibiotic-resistance genes in the plasmid metagenome of wastewater treatment plant bacteria showing reduced susceptibility to selected antibiotics. *Microbiology,* 155, 2306–2319.

Tacão, M., Correia, A. & Henriques, I. (2012) Resistance to broad-spectrum antibiotics in aquatic systems: anthropogenic activities modulate the dissemination of *bla*CTX-M-like genes. *Appl. Environ. Microbiol.,* 78, 4134–4140.

Tausova, D., Dolejska, M., Cizek, A., Hanusova, L., Hrusakova, J., Svoboda, O., Camlik, G. & Literak, I. (2012) *Escherichia coli* with extended-spectrum β-lactamase and plasmid-mediated quinolone resistance genes in great cormorants and mallards in Central Europe. *J. Antimicrob. Chemother.,* 67, 1103–1107.

Udikovik-Kolic, N., Wichmann, F., Broderick, N. A. & Handelsman, J. (2014) Bloom of resident antibiotic-resistant bacteria in soil following manure fertilization. *Proc. Natl. Acad. Sci. USA,* 111, 15202–15207.

Vaca-Pacheco, S., Miranda, R. & Cervantes, C. (1995) Inorganic-ion resistance by bacteria isolated from a Mexico City freeway. *Antonie Van Leeuwenhoek,* 67, 333–337.

Veldman, K., Van Tulden, P., Kant, A., Testerink, J. & Mevius, D. (2013) Characteristics of cefotaxime-resistant *Escherichia coli* from wild birds in The Netherlands. *Appl. Environ. Microbiol.,* 79, 7556–7561.

Voolaid, V., Tenson, T. & Kisand, V. (2013) *Aeromonas* and *Pseudomonas* species carriers of *ampC* FOX genes in aquatic environments. *Appl. Environ. Microbiol.,* 79, 1055–1057.

Walsh, T. R., Weeks, J., Livermore, D. M. & Toleman, M. A. (2011) Dissemination of NDM-1 positive bacteria in the New Delhi environment and its implications for human health: an environmental point prevalence study. *Lancet Infect. Dis.,* 11, 355–362.

Wang, N., Yang, X., Jiao, S., Zhang, J., Ye, B. & Gao, S. (2014) Sulfonamide-resistant bacteria and their resistance genes in soils fertilized with manures from Jiangsu Province, Southeastern China. *PLoS One*, 9, e112626.

Watkinson, A. J., Micalizzi, G. B., Graham, G. M., Bates, J. B. & Costanzo, S. D. (2007) Antibiotic-resistance *Escherichia coli* in wastewaters, surface waters, and oysters from an urban riverine system. *Appl. Environ. Microbiol.*, 73, 5667–5670.

Wichmann, F., Udikovik-Kolic, N., Andrew, S. & Handelsman, J. (2014) Diverse antibiotic resistance genes in dairy cow manure. *mBio*, 5, e01017-13.

Wirtz, V. J., Dreser, A. & Gonzales, R. (2010) Trends in antibiotic utilization in eight Latin American countries, 1997–2007. *Rev. Panam. Salud Publica*, 27, 219–225.

Yang, Y., Li, B., Zou, S., Fang, H. P. & Zhang, T. (2014) Fate of antibiotic resistance genes in sewage treatment plant revealed by metagenomic approach. *Water Res.*, 62, 97–106.

Yergeau, E., Lawrence, J. R., Waiser, M. J., Korber, D. R. & Greer, C. W. (2010) Metatranscriptomic analysis of the response of river biofilms to pharmaceutical products, using anonymous DNA microarrays. *Appl. Environ. Microbiol.*, 76, 5432–5439.

Zhang, X., Li, Y., Liu, B., Wang, J., Feng, C., Gao, M. & Wang, L. (2014) Prevalence of veterinary antibiotics and antibiotic-resistant *Escherichia coli* in the surface water of a livestock production region in northern China. *PLoS One*, 9, e111026.

Zhang, Y., Marrs, C. F., Simon, C. & Xi, C. (2009) Wastewater treatment contributes to selective increase of antibiotic resistance among *Acinetobacter* spp. *Sci. Total Environ.*, 407, 3702–3706.

Zhu, Y. G., Johnson, T. A., Su, J. Q., Qiao, M., Guo, G. X., Stedtfeld, R. D., Hashsham, S. A. & Tiedje, J. M. (2013) Diverse and abundant antibiotic resistance genes in Chinese swine farms. *Proc. Natl. Acad. Sci. USA*, 110, 3435–3440.

Zurek, L. & Ghosh, A. (2014) Insects represent a link between food animal farms and the urban environment for antibiotic resistance traits. *Appl. Environ. Microbiol.*, 80, 3562–3567.

Zurfluh, K., Abgottspon, H., Hächler, H., Nüesch-Inderbinen, M. & Stephan, R. (2014) Quinolone resistance mechanisms among extended-spectrum beta-lactamase (ESBL) producing *Escherichia coli* isolated from rivers and lakes in Switzerland. *PLoS One*, 9, e95864.

Zurfluh, K., Hächler, H., Nüesch-Inderbinen, M. & Stephan, R. (2013) Characteristics of extended-spectrum β-lactamase- and carbapenemase-producing Enterobacteriaceae isolates from rivers and lakes in Switzerland. *Appl. Environ. Microbiol.*, 79, 3021–3026.

Chapter 5

Impact of antibiotics and resistance in non-clinical settings

While conducting, but especially while writing papers about antibiotic resistance in the environment, it is very tempting to overstate the relevance of the findings, by asserting the grave danger of such resistant bacteria or resistance genes to find their way to humans, causing incurable infections. Indeed, the likelihood of such an event is very low, particularly compared to the immediate risks derived from the routine abuse of antibiotics in clinical settings: by simply staying in a hospital for a few days, the probability of acquiring a nosocomial infection is in the two-digit percent. However, it is very important to keep in mind that, when dealing with bacterial populations, that are amazingly large and diverse, and that multiply at great speed, even events considered to be very rare, inevitably do happen. This was very elegantly demonstrated by simple math: even if considering an incredibly low probability of an HGT event to occur, in the order of 10^{-24}, taking the bacterial population of the soil, and the surface planted with genetically modified corn, about 5,000 recombinants would be expected to exist; but about 500 million tons of soil would have to be analyzed to detect one (Heinemann and Traavik, 2004). The currently common finding of CTX-M beta-lactamases and *qnr* genes in clinical isolates, is the result of a sum of very unlikely events: the mobilization of said resistance determinants, from environmental, non-clinically-relevant species, into commensal and pathogenic bacteria; and the acquisition of the latter by humans, in amounts and conditions that resulted in an actual infection. And while the individual risk of a person to become infected by an environmental resistant bacteria may be negligible, the accumulated morbidity and mortality (and healthcare-related economic costs) of infections caused by CTX-M beta-lactamases worldwide, may illustrate just how relevant the issue of resistance in the environment can get.

Furthermore, the accelerated invasion and destruction, of environments that were until recently essentially free from human influence, along with climate change and increased globalization, have put together microorganisms, animals and humans that did not use to be. Lyme disease, hantavirus and *Vibrio* infections are but a few examples of this (Cohen, 2000). And to top it all off, microbes are traveling in old and new ways, extending the reach of any and all of these problems worldwide. Resistant organisms being selected by the agricultural use of antibiotics in one country can travel along water streams and airborne animals into another; a wild animal captured from a polluted rural area can end up as a pet in a distant city, and share its resistant microbiota with people and other animals; foodstuff coming from countries where rampant abuse of antibiotics is common can be exported all over the world; unthinkable vehicles, such as ballast water from ships, can carry huge amounts of pathogenic bacteria between countries (Ruiz et al., 2000). All these is to say that we simply do not have the tools for assessing the risks of the continuing abuse of antibiotics, and of the massive release of antibiotics and bacteria into the environment; nor do we have an idea of the potential consequences of having pathogenic, multi-resistant bacteria in remote rural environments and in wildlife. Unfortunately, as so many times in the past (*e.g.*, radiation and radioactive material, asbestos, PCBs, halocarbons, tobacco, "mad cow disease") and despite early warnings, risk assessment "experts" swiftly dismiss alarm calls, endangering people and environments to protect business. "All of this has happened before, and it will all happen again."

5.1   IMMEDIATE RISKS

While it is so far impossible to assess the actual probability, there are some events that have already been documented, so that these and similar situations can be labeled as "immediate risks" (Figure 5.1; in this chapter, the whole text would function as "figure legends"; hence figures would only be numbered without further legend, except for the last one). Antibiotics are used, at inhibitory concentrations (the hand bomb

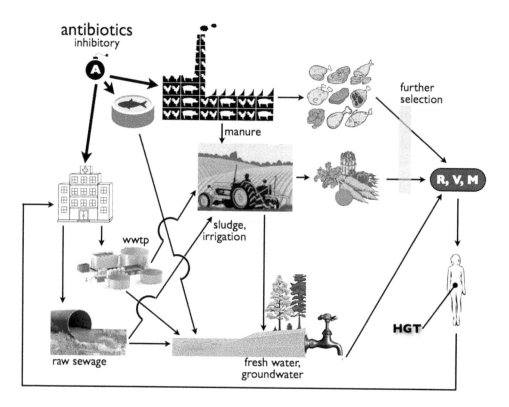

Figure 5.1 Antibiotic resistance in the environment: immediate risks.

marked with A) for agricultural and clinical purposes; kinds, amounts and routes were described in Chapter 3. Within humans and animals receiving those antibiotics, as well as within affected micro-environments (*e.g.*, farms, ponds, and even hospitals), bacteria (grey rounded rectangle, right) carrying resistance (R) determinants, often along virulence (V) and/or mobility (M) genes, are selected. These organisms can reach urban or rural human populations through several different routes: meat and other animal products are often contaminated with resistant bacteria; produce may also contain resistant bacteria if manure or sludge from wastewater treatment plants (wwtp) were used as fertilizers, and/or if wwtp effluent or raw sewage were used for irrigation. Although many animal products are cooked, and resistant bacteria are mostly eliminated, people handling raw meat, as well as household instruments and surfaces that come in contact with raw meat, may become contaminated with such microbes; produce is often eaten raw, and disinfection procedures may not be enough to eliminate bacterial contamination. While there are many examples of the impact of the agricultural use of antibiotics as raising resistance rates in clinical isolates, a single one may suffice: fluoroquinolone resistance, particularly in *Campylobacter*, a bird commensal, in countries using such fluoroquinolones agriculturally (McEwen, 2012). This phenomenon may not be limited to *Campylobacter*: Asian countries, massive producers of chicken meat and massive agricultural users of fluoroquinolones, had the highest fluoroquinolone resistance rates among unrelated pathogens, such as pneumococci and gonococci (Huang and Hsueh, 2010).

Microbes deliberately added to foodstuff, as starter cultures, probiotics, or as biopreservation strategies, may further increase the load or resistance genes; and food processing and preservation methods (*e.g.*, refrigeration, salt or acid addition) may select or induce resistance phenotypes (Verraes et al., 2013). Effluents from wwtp, as well as raw sewage carrying resistant bacteria, are released in water bodies which are, in turn, sources of drinking water; resistant bacteria have been found in tap water, even after chlorination procedures (heavy-metal resistance or tolerance, which is often genetically linked to antibiotic resistance determinants, has been recognized as a common feature of bacteria in drinking water distribution systems (Ford, 1994)). While the main concern would be for resistant enteropathogens to be acquired through these routes, other resistant bacteria may simply colonize people eating contaminated food and, through HGT, resistance genes be transferred to other kinds of pathogens, capable of causing extraintestinal infections.

5.2 "HOTSPOTS" FOR RESISTANCE SELECTION AND HGT-MEDIATED REARRANGEMENTS (FIGURE 5.2)

Intensive antibiotic usage, at hospitals and farms, exert equally intensive selective pressure upon human and animal pathogens, commensals, and bacteria living in such environments (*e.g.*, rumor has it that the droplets of concentrated antibiotic solution that are expelled along with the bubbles while purging a syringe, amount to ~30 L/year of such solutions dropped in the floor of large hospitals). Hospital sewage contains excreted bacteria from antibiotic-treated patients, which may be commensal or pathogenic, but are almost always resistant to at least a single drug; along with excreted antibiotics (available data were reviewed in Chapters 3 and 4). Although most

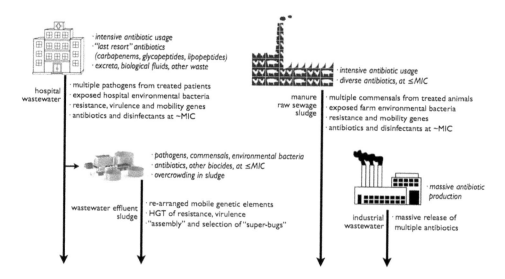

Figure 5.2 "Hotspots" for resistance selection and HGT-mediated rearrangements.

of the antibiotics used clinically are taken by outpatients, concentrations of bacteria and antibiotics in house sewage are much more diluted; furthermore, antibiotics used in hospitals are more frequently of the "last resort" kind: carbapenems, glycopeptides and lipopeptides, along with other drugs not so commonly used in ambulatory patients, such as aminoglycosides, third-to-fifth generation cephalosporins and oxazolidinones. Therefore, bacteria in hospital sewage are more likely to be virulent and resistant; and antibiotics to be of a more dangerous nature and at higher concentrations. While it has not been documented, it is likely that the immediate effluent of hospital sewage can provide conditions for intensive HGT and antibiotic selection, perhaps only limited by the rapid liquid flow: biofilms at sewage pipes should be teeming with multi-resistant organisms, exchanging genes at high rates, if we are to consider the biofilm environment as a HGT hotspot (Sørensen et al., 2005); or planktonic organisms may be the ones more prone to carry mobile elements of all sorts, as seems to happen with marine vibrios (Xue et al., 2015). In non-developed countries, most of this wastewater is directly dumped into the environment; where wastewater treatment plants are available, bacteria from hospitals and houses, along with urban environmental organisms carried by runoff into sewers, are all concentrated in a sludge that also contains antibiotics, disinfectants and other toxic compounds. As discussed in the previous chapter, these conditions seem to foster gene rearrangements, HGT and intensive selection, resulting in the gathering of resistance determinants in mobile genetic elements, their mobilization among different bacterial hosts, and the enrichment of these newly formed "superbugs" within the sludge. Other mostly unknown interactions of fecal bacteria with other microbes in wastewater treatment plant sludges, involve protozoans: the presence of enterohemorragic *E. coli* (EHEC), producing Stx toxins -believed to be primarily a bacterial defense against protozoan predators, changes the composition of the protozoan population of wastewater sludge (Li et al., 2015). The enhanced survival or resistant EHEC caused by antibiotics and related biocides, may further modify the number and nature of the overall bacterial load in sludges. Treated wastewater is in turn released into water bodies or used for irrigation purposes; and sludges are used as fertilizers upon agricultural soils. Several resistance and mobility genes have been detected in water environments, mostly linked to wastewater (Lupo et al., 2012).

The use of antibiotics on food animals represents, as previously reviewed, the main form of antibiotic usage: many different antibiotics are used at therapeutic and sub-therapeutic dosages upon many millions of animals. Complex and unpredictable events do happen to the microbiota of such animals (*e.g.*, antibiotics in animal food induce prophages in the intestinal microbiome of animals (Allen et al., 2012), modifying the microbiota itself, and potentially fostering HGT through transduction); in the end, antibiotics and bacteria (both, commensals and pathogens) are excreted by treated animals. Waste from factory farming, including excreta, carcasses, food, bedding, etc., all containing antibiotics and bacteria, is directly applied to agricultural soils, mostly as solid manure; sewage and sludges may undergo some biological treatments in lagoons; and manure is sometimes composted; while such treatments may reduce the prevalence of some resistance determinants, others can persist and even increase (Pruden et al., 2013). Within solid and liquid manure coexist released bacteria from medicated animals, environmental bacteria, and inhibitory concentrations of antibiotics, disinfectants and heavy metals, among other biocides; this conditions sound like a breeding media for genetic rearrangements and HGT, the same as sludge at wastewater treatment plants. Gene rearrangements within these conditions may even produce new resistance determinants, as has been shown for the mosaicism found in some *tet* genes from animal and human commensals (Aminov and Mackie, 2007). After being applied to agricultural soils, antibiotics and microbes tend to dilute; but when still present at inhibitory concentrations, in addition to the expected selection of resistance determinants, antibiotics modify the composition of microbial communities, favoring some bacterial taxa (Xiong et al., 2015). As to their influence on water bodies, resistance genes being selected by the use of antibiotics in farms and present in manure lagoons, can be detected in groundwater (Koike et al., 2007).

Pharmaceutical factories from both, developed and non-developed countries, seem to release copious amounts of antibiotics in their wastewater; in the former, such wastewater undergoes treatment, while in the latter it is mostly released directly into water bodies. Total quantities and concentrations vary widely, as discussed in Chapter 3. In any event, before being diluted, these antibiotics end up exerting a strong selective pressure, either upon sewage bacteria in treatment plants and water bodies; or upon water or soil environmental bacteria. Although the evidence is scarce, some reports have documented the increased prevalence of resistant bacteria downstream of such factories (Pruden et al., 2013).

5.3 ECOLOGICAL IMPACT UPON SOIL BACTERIA (FIGURE 5.3)

Soil microbiota (bottom right) is considered a well-structured, resilient community, that include antibiotic producers (carrying antibiotic biosynthetic genes, A [in old-fashion font], and corresponding resistance genes, R [also in old-fashion font]), along

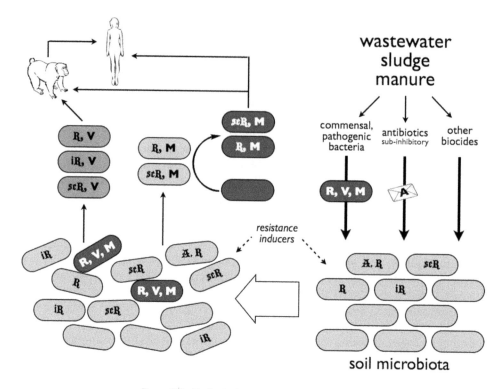

Figure 5.3 Ecological impact upon soil bacteria.

with many kinds of other bacteria with plain resistance (R), intrinsic resistance (iR) or sub-clinical resistance (scR). When receiving wastewater for irrigation purposes, and/or sludge or manure as fertilizers, an important bacterial load is added, mostly composed by human or animal commensal or pathogenic bacteria, carrying resistance, virulence, and/or mobility genes (R, V, and M, in modern font). Antibiotics, mostly in sub-inhibitory concentrations (the envelope marked with A, representing antibiotics at concentrations acting as signaling molecules, instead of as selective pressure; our actual ability to measure antibiotic concentrations in environmental samples has been questioned (Keen and Patrick, 2013), but most authors agree in that amounts released to, and persisting in the environments are in the far sub-inhibitory range, as discussed in Chapter 3), are also added, along with other biocide compounds, such as metals (Cu, Zn, As) commonly used in animal feeds instead of antibiotics, with co-selecting capability and more persistent in agricultural soils (Pruden et al., 2013). Resulting from this exposure, antibiotics may disturb the microbiota, by affecting cell physiology and/or selecting for resistant organisms: sub-MIC antibiotics exert a number of effects as signaling molecules, directly affecting translation, transcription (5–10% of bacterial transcripts are affected by antibiotics at subinhibitory concentrations), and replication, and also acting as quorum signals (Aminov, 2009, Yim et al., 2007); significant changes in metabolic activity have been observed in bacteria from river sediments, up- and down-stream of wastewater discharges (Janzon et al., 2012); and sub-MIC antibiotics (*e.g.*, 100 ng/L ciprofloxacin, not rare in wastewater effluents) may "provide a small but measurable selective advantage to resistant bacterium" (Larsson, 2014). It has been suggested that antibiotics may disturb microbial functions that are crucial for plant growth, such as nitrogen fixation (Larsson, 2014). Other, non-antibiotic agents, may affect the soil microbiota not by exerting a direct selective pressure, but by inducing mechanisms of adaptive resistance. From the agricultural-environmental point of view, some herbicides, such as dicamba, glyphosate and paraquat, were mentioned in previous chapters as modifiers of the response to antibiotics; but there is a large list of bacterial responses to environmental stress, including oxidative, nitrosative, envelope-, temperature-, and nutrient-related, that can elicit an antibiotic resistance phenotype (Poole, 2012). Such conditions may exist, or be further added by human influence, and affect environmental bacteria and newcomers alike. Transient, inducible resistance may enable the survival of bacteria exposed to also transiently to inhibitory concentrations of antibiotics (*e.g.*, contaminated wastewater release, or manure application), thus preferentially allowing the survival of organisms carrying the inducible trait and under the influence of the induction stimuli.

Within the iR category, efflux pumps play a significant role; efflux pumps are very common in soil bacteria, especially among those living in close proximity to plants. The prefix "multidrug" is often associated to their name (as in MDR pumps, for "multidrug resistance"), conveying the wrong notion that these are mainly antibiotic resistance mechanisms, and pushing them into resistance gene databases. However, these pumps are only a sort of unspecific vomit system for bacteria, allowing for a low-level resistance to a variety of xenobiotics. In non-clinical environments, these pumps have been involved in plant pathogenesis, enabling bacteria to avoid the toxic effects of plant-derived antimicrobial compounds; they may also play a role in regulating the responses to quorum-sensing molecules and other cell-to-cell chemical cross-talk (Martinez et al., 2009). The prospect of having such complex, tightly-regulated machineries mobilized and expressed as full-resistance phenotypes by clinically-relevant bacteria is rather slim. However, as antibiotics, metal ions and disinfectants are all extruded by these pumps, such agents, particularly at sub- or borderline-inhibitory concentrations, can exert a selective pressure favoring bacteria that overexpress them, including constitutive mutants. This would be yet another way for chemical pollution to modify soil microbiotas, potentially affecting their ability to communicate internally, and to infect larger organisms.

A brief parenthesis must be devoted here to biofilms. When discussing biofilms in the context of antibiotic resistance, the main issue is the resistance or rather persistence of these bacterial communities; many reasons and models to explain this have been proposed (Gilbert et al., 2007), especially since the great majority of infectious episodes involve biofilm formation. Importantly, most of the research on biofilms use very artificial conditions, such as attachment to the glass or plastic walls of tubes containing liquid broth; and, especially, exploring the behavior of biofilms formed by individual strains, instead of the most likely natural scenario of multi-species communities. In addition to the role of biofilms in pathogenesis and antibiotic therapeutic failure, they are relevant to the issue at hand as they also exist in the environment, from sewage pipes and wastewater plant sludges, to river beds, to soils. Biofilms can be construed as the ideal cooperative microbial community: biofilms of *Bacillus subtilis*, a soil bacterium, develop a metabolic co-dependence between the outer and inner layers of the biofilm, with periodic oscillation in growth rate that compensate the advantages (nutrient availability for the outer layers, protection for the inner ones) and disadvantages of different subgroups of the same bacterial community (Liu et al., 2015). But when different strains of the same species are mixed, biofilms seem to be something entirely different: by mixing strains of *P. aeruginosa*, biofilm formation is enhanced, but is rather the result of fierce competition, with one strain killing off the other (by chemical warfare: pyocin production), and biofilm formation being a response to cellular damage. In this scenario, sub-inhibitory antibiotics (ciprofloxacin, rifampicin, tetracycline) induce the formation of biofilms, perhaps as a defense mechanisms towards a detected threat from a foreign, antibiotic-producing strain (Oliveira et al., 2015). While this limited experience with *P. aeruginosa* not the typical soil bacteria, returns us to the antibiotics as chemical warfare notion, it is clear that, whatever the case, sub-MIC antibiotics in the environment can have profound consequences on microbial ecology.

Soil microbiota would also be modified by adding organisms that are not typical of such environments, as are most fecal bacteria. Commensal and pathogenic bacteria may not survive long in the soil, as it is a hostile media with limited nutrients and adverse physical-chemical conditions; but perhaps long enough to mobilize genes into soil bacteria. The increased presence of sulfonamide resistance genes and mobile elements in soil bacteria after adding manure, for instance, have been documented (Heuer et al., 2012). Such gene mobilization may further allow soil bacteria to act as donors of resistance genes, that are then acquired by previously susceptible commensals or pathogens; or may enable virulence genes to be acquired by resistant soil bacteria, which then can infect wildlife or people. This is not to say that undisturbed soil microbiota is not used to HGT: the rhizosphere, for instance, has been recognized as a hotspot for gene mobilization (Sørensen et al., 2005). However, mobility genes from commensals and pathogens may be just what is needed to break the apparent genetic isolation of bacteria in the soil environment (Forsberg et al., 2014), and to allow the gene flux from the "soil microbiome" into clinically-relevant microbes; this would be discussed below.

5.4 POTENTIAL ECOLOGICAL AND CLINICAL IMPACT OF RESISTANCE IN WILDLIFE (FIGURE 5.4)

Antibiotics (and other biocides capable of co-selecting antibiotic resistance, such as disinfectants, heavy-metals, non-antibiotic drugs, etc.; continuous lines) and bacteria resistant to antibiotics (and/or to other agents, but enabling co-selection; dotted lines) can reach terrestrial wildlife in a number of ways: manure and sludge from factory farming is probably the main source, as it is applied in massive quantities in agricultural soils and/or released to water bodies; raw sewage and wastewater treatment plant effluents represent the main input from urban settings, while landfills may contribute minimally. Airborne animals, such as birds and flying insects, can provide a long-range airlift for resistant bacteria, which they often carry, from all of the options at the left. Wild animals living in close proximity to human settlements may forage through agricultural soils, becoming directly exposed to these sources; those living far enough may

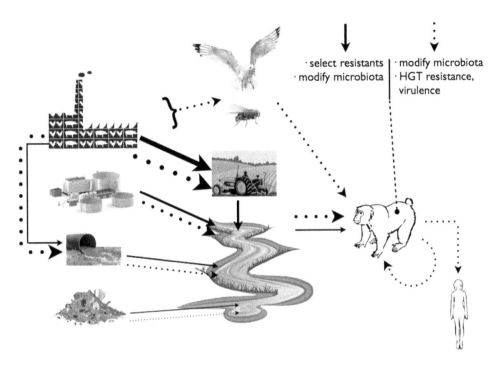

Figure 5.4 Potential ecological and clinical impact of resistance in wildlife.

still receive some antibiotics and resistant bacteria, mostly carried by water streams and airborne animals. Water seems to play a significant role in the dissemination of resistance: wastewater is constantly released to connected water bodies, and used for many purposes; even coastal waters may contain large amounts of enteropathogens – likely coming from sewage, some of them carrying resistance genes (Finley et al., 2013). Uptaken antibiotics are mostly in the sub-inhibitory concentrations; however, as discussed previously, even such low concentrations may select for resistant bacteria. This, and the potential signaling effects of sub-MIC antibiotics, may modify the microbiota of affected animals. As has been recently documented for the microbiota of humans and other mammals, such changes may induce physiological (Claesson et al., 2012, Gareau et al., 2010, Maslowski et al., 2009, Ottman et al., 2012), metabolic (Claesson et al., 2012, Claus et al., 2011, Payne et al., 2011, Qin et al., 2012, Turnbaugh et al., 2009) or even behavioral (Diaz Heijtz et al., 2011) modifications in wildlife of which we know nothing about.

Wildlife would also uptake bacteria released from these sources, including foreign pathogens and commensals, which may also be resistant to antibiotics and other biocides, and carry an unusual load of mobile genetic elements. The acquisition and establishment of such bacteria may also modify the microbiota of the affected animals; and the presence of mobile elements may foster HGT between resident and recently-acquired microbes. As a result, newly-resistant and/or newly-pathogenic bacteria may be transmitted to other animals, causing an outbreak; and/or be transmitted to humans. From this perspective, the guts of wild animals receiving resistant bacteria from polluted environments may be conceived as additional hotspots for the brewing of resistant commensals or pathogens for humans: animals would more likely host incoming human bacteria than open environments, their microbiota would be phylogenetically related, allowing for easy and successful HGT; and the lines separating human and animal pathogens are thin and blurry.

5.5 ICEBERG TIPS, KNOWN KNOWNS, AND UNKNOWN UNKNOWNS (FIGURE 5.5)

A -not at scale- representation of all bacteria (open triangle), including culturable ones (shaded triangle) and clinically-relevant ones (*i.e.*, those clearly causing infectious processes, and also members of the human microbiota that are involved in physiological and metabolic homeostasis; diamond shape). Some bacteria carry resistance genes

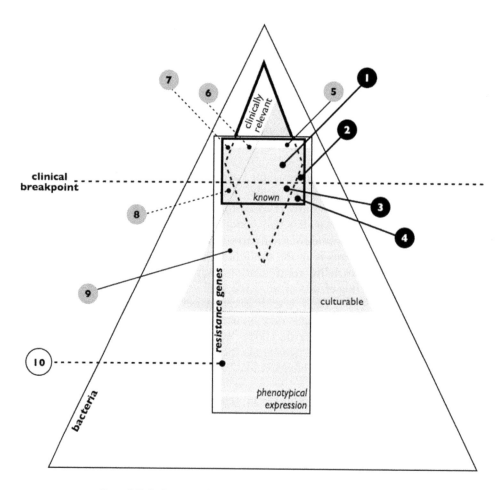

Figure 5.5 Iceberg tips, known knowns, and unknown unknowns.

(open rectangle, thin line), including known ones (open rectangle, thick line) with phenotypic expression (shaded rectangle). A dashed line indicate the breakpoint for resistance from the clinical point of view, with full-resistant ones above the dashed line. Most information about resistance comes from clinically-relevant (1) and non-relevant (2) culturable bacteria, carrying known resistance genes with phenotypic expression above clinical breakpoints (*e.g.*, Enterobacteriaceae with ESBLs, *Streptomyces* with aminoglycoside-modifying enzymes, respectively). Some clinically-relevant (3) and non-relevant (4) culturable bacteria carry known resistance genes whose expression yields a low-level, below-breakpoint resistance (*e.g.*, *qnr* genes conferring low-level quinolone resistance in Enterobacteriaceae and aquatic bacteria, respectively). There is some information about (5) known genes without phenotypic expression in clinically-relevant, culturable bacteria (*e.g.*, *tet*(X) in *Bacteroides*). Then there is molecular evidence (provided by PCR-based or metagenomic assays) of known resistance genes in unculturable bacteria both, of putative clinical-relevance (6) or not (7), for which the actual expression in the original hosts cannot be tested; there is also evidence of known, low-level resistance genes in non-clinical, unculturable bacteria (8). Some culturable bacteria without clinical relevance (9) carry previously unknown genes whose expression yield a low-level resistance. But there could be (10) a vast variety of unculturable, non-clinically-relevant bacteria, carrying unknown resistance genes that could not even confer actual resistance in their natural hosts, but do so when mobilized and expressed in pathogens. All of the above sets are in continuous flow, and boundaries are weak, primarily because of HGT. But it may be useful to look at the instant picture in order to understand the risks posed by each set.

Known, culturable, pathogenic organisms displaying actual resistance (*i.e.*, above clinical breakpoints; 1) are the only ones we know for sure that would be related to clinical failure of antibiotic treatment; those expressing only low-level resistance (3) may conduce to clinical failure in particular conditions (*e.g.*, when infecting tissues not reached by inhibitory concentrations of antibiotics, or when antibiotics are not used in adequate doses), and are likely to gain, through mutation, HGT or adaptive responses, a full-resistance phenotype during treatment. Culturable but non-pathogenic bacteria, carrying high- or low-level known resistance genes (2 and 4), represent the known examples of resistance traits from the environment that have "jumped" to pathogens; and may continue to contribute such determinants to the gene pool available to clinically-relevant microbes. Our certainties end there.

There is a number of clinically-relevant bacteria that we have been unable to culture. It is difficult to pinpoint a clear example of an infectious disease caused by an unculturable bacteria (nanobacteria supposedly linked to urinary stone formation (Kajander and Çiftçioglu, 1998), which may actually be resistant to antibiotics (Sardarabadi et al., 2014) come to mind), and much more of one carrying resistance genes; but the probability certainly exists. For instance, before 1982, when Marshall and Warren linked *H. pylori* to gastric ulcers, such illness was not considered infectious in nature, antibiotics were not used against it, and resistance was not an issue. This however does not represent an immediate health jeopardy, as none of such clinical conditions are currently considered an infection, hence antibiotics are not used for treatment (and antibiotic failure due to resistance is not an option). But, should one of such previously unknown infections is recognized, resistance would become an issue. On the other hand, the great majority of the human microbiota have not been cultured, and qualify as "clinically-relevant"; metagenomic approaches have detected known and previously unknown resistance genes (6), of whose actual expression we know little about, but that may have an impact on the whole microbiome when the human host is exposed to antibiotics. Should, for example, a bacterial genus as *Prevotella* somehow linked to obesity (Arumugam et al., 2011), is to acquire resistance determinants, antibiotic treatments may foster the prevalence of such organisms and, in turn, the risk towards obesity. For now, this is merely hypothetical, but the health repercussions of such an unexpected result of antibiotic presence and resistance may be significant.

Culturable environmental bacteria, such as *Streptomyces*, seem to carry a variety of resistance genes, some of them not previously known (and, actually, that cannot be clearly defined as "resistance genes"; 9). Applying functional metagenomics to culturable soil proteobacteria, a number of resistance genes previously found in pathogens (*e.g.*, *tet* genes, aminoglycoside modifying enzymes, *sul1*) were detected, but 54% of the 110 putative resistance genes were not previously known (Forsberg et al., 2012). These, however, included an open reading frame that conferred "resistance" to cycloserine (a second-line drug used only against tuberculosis, obtained from *Streptomyces*) to the *E. coli* recipient; and 55 beta-lactamases, mostly of class C (*i.e.*, typical chromosomal ones); the actual relevance of such "resistances" is doubtful. And, again, metagenomic approaches have provided evidence for the presence of genes in unculturable environmental bacteria that could be construed as "resistance" ones. Some of the limitations of such techniques were mentioned in Chapter 1, and also in a recent review (Mullany, 2014); nevertheless, these are the only available approaches to start making an educated guess regarding the magnitude and potential impact of the "resistome".

5.6 ALL THIS IS VERY INTERESTING BUT...

What is, in the end, the relevance of looking and finding antibiotics and antibiotic resistance in the environment? As has been reviewed here and by others (Bernier and Surette, 2013), antibiotics in the environment may act, as signaling molecules at low concentrations, potentially disturbing environmental microbiotas; or as selective pressures, at high concentrations, favoring resistance and other linked traits (*e.g.*, resistance to other toxic agents, gene mobility, etc.). The precise boundaries between one effect and the other are not clear, as it is well-known that even a sub-MIC presence can increase the prevalence of resistant bacteria, and maintain mobile elements carrying resistance determinants: concentrations 140-fold below the MIC of antibiotics and heavy metals were enough to stably keep a 220-kbp plasmid encoding and ESBL and resistance to tetracycline, aminoglycosides, macrolides, sulfonamides, trimethoprim, ciprofloxacin, silver, copper and arsenic (Gullberg et al., 2014). Also, sub-MIC effects differ from one species to another: while aminoglycosides, tetracyclines and chloramphenicol do not induce an SOS response in *E. coli*, they do, along with fluoroquinolones in *V. cholerae* (Baharoglu and Mazel, 2011). Furthermore, "resistance", aside from the breakpoints used to guide the clinical use of antibiotics, is more of a continuum than a black-and-white picture. All antibiotics, at low or high concentrations, that have been detected in the environment, are human-made pollutants, as naturally-produced antibiotics exist in amounts so little that cannot be detected by current technologies.

On the other hand, resistance genes can be viewed from two main perspectives: (a) as pollutants themselves, having being selected by the human use of antibiotics, clinical or otherwise, that are copiously released into the environment in many different ways; and (b) as ancient genes, carried by environmental bacteria, some of them protecting antibiotic-producing organisms, some protecting the annoyed neighbors of the latter, but most of them of unknown biological function – some even without any relation to xenobiotic protection, only revealed as "resistance" by the artifacts of metagenomics. Antibiotics, antibiotic resistance genes, along with other agents and traits, coexist in different environments; the results of such coexistence is simply unpredictable, given the enormous set of variables at play, and their dynamic nature. Some of the possible consequences of unleashing so many powerful elements in so large quantities, have been outlined all along this text. There is one, however, that has been analyzed recently, and upon which a strong controversy exists: the actual likelihood of ancient resistance genes to reach pathogenic bacteria, adding to the current resistance crisis in the clinical setting. The well-known examples of this actually happening (CTX-M beta-lactamases and *qnr* genes) only serve as proof-of-concept, but do not clarify if they were the exceptions to a rule ("HGT [do not] effectively decouples resistomes from phylogeny" (Forsberg et al., 2014)) or, if not, what are the conditions necessary for an ancient resistance gene to become a present-day scourge. The identification of such conditions is crucial to, among other things, extract real knowledge from the wealth of data that metagenomic approaches are yielding. All other considerations aside, the ability to timely detect an impending public health risk when looking and resistance environmental surveillances, would be the most important contribution that this line of work is expected to provide.

In the process of transferring ancient genes to current pathogens, Martínez identified some issues aptly named "bottlenecks" (Martínez, 2012). He identifies the need for donors (soil bacteria) and recipients (commensals and pathogens, or a possible intermediate group, organisms "more prone to exchange genetic material") to share the same habitat; then, the delicate balance between fitness cost of the acquired gene, and the presence of a selective pressure, to allow for its successful establishment: the higher the fitness cost, the higher the need for a strong selective pressure, and vice versa. Also, there is competition between a possible previous resistant determinant and the newcomer within the affected microbial population; finally, a number of further issues, such as co-selection, by linked genes or cross-resistance, and the inherent stability of the mobile element serving as vehicle – most likely a plasmid. If these bottlenecks are not considered, the sole presence of a "resistance" gene in a "resistome", such as an unspecific, chromosomally-mediated efflux pump, would only indicate the presence of the bacterial taxa they are linked to (*e.g.*, *E. coli*, if *acrAB* are detected; or *P. aeruginosa*, if *mexXY*), but not real resistance prevalences, nor the actual likelihood of such genes to reach the commensal-pathogen bacterial community. To start trying to assess such risks, it is crucial to identify said bottlenecks, and to incorporate this knowledge into resistance gene databases. Martínez *et al.* suggest some "alert levels" (RESCon, "resistance readiness condition", likely from the typical humor of F. Baquero): 1 for genes known to cause antibiotic treatment failure and that reside on mobile elements; 2 for novel genes that encode antibiotic inactivating enzymes that, again, are carried by mobile elements; 3 for genes that confer resistance to novel antibiotics; 4 for genes mediating resistance by unknown mechanisms, towards antibiotics whose clinical efficacy is already affected by other known resistance genes; 5 for new variations of known resistance mechanisms; 6 for genes predicted to confer resistance and known to reside in mobile elements; and 7 for those predicted to confer resistance but not linked to a mobile element (Martínez et al., 2015b). Immediately after publication, some authors disagreed on the proposed order: RESCon 1 category for known genes in mobile elements is, in their opinion, unwarranted, as for such a gene found in the environment their risk of "returning" to clinical settings would be low, lower at least that the direct mobilization within clinical settings; there were also other minor issues raised (Bengtsson-Palme and Larsson, 2015). Martínez *et al.* counter argued that their proposed scale was not intended only for use in environmental studies, but on all metagenomic results; and that the mobilization, instead of a single-step process of environmental-to-pathogen transfer, supposedly rare, could involve several intermediates, raising the risks posed by the finding of a known, mobilizable resistance gene in an environmental survey (Martínez et al., 2015a). Both groups seem to agree on the particular importance to identify hotspots, *i.e.*, sites where many different bacteria, carrying many different resistance genes in many different mobile elements, converge under the selective pressure of high concentrations of many different antibiotics. Under such conditions, said bottlenecks would loosen up; therefore, RESCon categories should change, depending on the source of a detected resistance gene. All these considerations may, indeed, allow us to gain some insight from metagenomic studies, other than the knowledge that there are some resistance genes out there. However, it is necessary to go a bit deeper.

The linkage of environmental resistance genes to mobile elements, is considered a crucial step for such a gene to end up causing health problems. This is highlighted in the RESCon scale mentioned before; and the finding of much fewer mobility genes in the vicinity of resistance genes in the environment – much less than in clinical isolates, was considered as further evidence for the lack of mobilization of resistance within soil microbiomes (Forsberg et al., 2014). However, these considerations fail to include three very important notions (Figure 5.6): (a) a huge variety of mobile elements are being continuously supplied to soils and waters; (b) mobile elements are, well, mobile; and (c) antibiotics and other agents, also released to the environment, do act, directly or indirectly, as inducers of gene mobility both, intra- (gene cassette mobility, transposition) and inter-cellularly (conjugation and transduction). It is therefore very likely that, in a sort of retrotransfer event, incoming commensal or pathogenic bacteria with the kind of complex, conjugative plasmids that are typical in wastewater treatment plant sludge, transfer such plasmids to environmental species, that become further donors of resistance after some minor genetic rearrangements, all under the inducing and perhaps even selecting activity of antibiotics.

Other, less likely but not impossible scenario, would be the conversion of a resistant, environmental bacteria, into a human or animal pathogen or commensal. In this case, the mobility of the resistance gene would not be necessary; actually, this would be the one circumstance where intrinsic resistance of environmental bacteria may become relevant. It would be important to identify those environmental bacteria that are closer to gain a commensal/pathogen lifestyle; and/or those individual genes or islands that may allow such a conversion. If all said elements converge in a single environment, it would become feasible for an environmental resistance gene to gain clinical relevance.

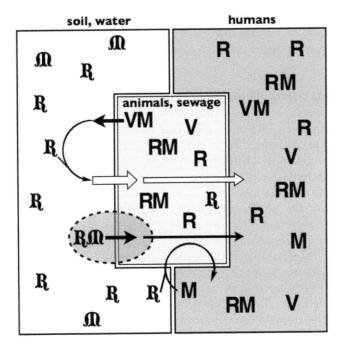

Figure 5.6 **Resistance and gene flow in the environment.** Freely based on the excellent figure at Box 2 of Martínez *et al.* paper (Martínez et al., 2015b), this one would aim at complementing it. Three somewhat separated microbiotas, human, animal/sewage, and soil/water are considered; antibiotic pressure upon each of them is represented as a gray shade: high upon human microbiota; variable albeit lower in average, in animals and sewage; rather low in soils and waters. Resistance (R) and mobility (M) genes are present in all three, while virulence (V) genes only in organisms from animals and human. Some are ancient (old style font) and some seem to be more recently concocted (modern font). Aside from the ecological impact of the release of antibiotics and bacteria on different environments, discussed all along this book, the remaining question is about the risk of having an ancient resistance gene mobilized into a human pathogen. Resistance and mobility genes exist in pristine environments, but they do not seem to be closely linked – as they do in clinical and wastewater isolates, suggesting a barrier for the actual mobilization of ancient resistance determinants to the human microbiota. However, should ancient resistance and mobility genes do concur, and under the inducing and selective influence of antibiotics (shaded, dashed outlined oval), it is possible that such a gene may travel directly or indirectly into the human microbiota. This may be just what happened to CTX-M beta-lactamases or *qnr* genes. Two further scenarios are proposed: a virulence-mobility combination being gained by a soil organism already carrying resistance determinants, which would enable the acquisition of the whole organism by humans (open arrows); and the "retrotransfer" of resistance, by means of a mobility gene being linked to an ancient resistance one, and the further mobilization of the combined element back to a human commensal or pathogenic bacterium.

Of course, this scenario would be less far-fetched if talking about the microbiota of wildlife.

The final conclusion would be that, in addition to further annotation of resistance genes in databases (*e.g.*, Gibson et al. (2015)), that may improve the interpretation of metagenomic results, it would be equally important to identify bottlenecks, hotspots, and the actual dynamics of HGT; and, possibly, to design specific experiments and databases for each major scenario (*i.e.*, human gut, wildlife feces, soils, sludges, etc.) instead of the current one-size-fits-all approach.

5.7 CONCLUDING REMARKS

Reviews and books on antibiotic resistance in the environment tend to include, at the end, a section or chapter devoted to regulatory issues: what to do to contain the release of antibiotics and the spread of resistance? As a citizen of a non-developed country, where the environmental regulatory framework, if any, is permanently bypassed by incompetence and corruption, this author puts little faith and effort into this issue. As most of the world population lives in non-developed countries, this reality is, sadly, the prevalent one; furthermore, after looking at the delayed, timid and incomplete responses from international organizations; and at the effective lobbying against national regulations in many developed countries, it seems only clear that non-developed ones do not hold the monopoly of incompetence and corruption. It is also evident that even well-meant risk-assessment efforts, upon which most regulations must rely, are far from fully understanding all variables at play, and from being capable of detecting extremely-rare events that can easily become very serious risks. As with many other environmental issues, such as climate change and animal extinction, regulatory efforts have only partial effects in rich countries, and fail dismally at the global scale.

There is, however, one common theme that has to be put forward: the pervasive effects of free-market theologies. From the clinical side, free-market notions have only allowed antibiotic abuse, especially of broad-spectrum antibiotics that are good for business as they can be used against many diseases; the abandonment of antibiotic R&D, as there are other more profitable avenues for pharmaceutical research; and the immoral notion of the need for "incentives", including higher prices, to have *big-pharma* back to the antibiotic business. From the agricultural arena, which is the main antibiotic abuser, the only reasons for antibiotic usage are of financial nature, most particularly the massive use of antibiotics for "growth promotion". Agricultural use of antibiotics, of *all* antibiotics – not only those without direct clinical use, must cease immediately, worldwide. This would prevent the further selection of resistant organisms within food animals, which in turn get into our foodstuff; and the release of antibiotics and resistant organisms in the many forms of waste these activities generate, that end up one way or another in the environment.

From the environmental perspective, degradation is an inevitable consequence of growing human populations and demands; it is only an extension of the second law of thermodynamics. However, to arbitrarily assign risks for each of the damages we do to the environment; and to define "tolerable risks" only based on their (perceived) financial repercussions, is shortsighted at best, and plainly immoral most of the time. Setting of maximum amounts of antibiotics and bacteria released by wastewater treatment plants, based on completely imaginary "risk-assessment" perceptions, is but a single example. It is certainly possible to prevent the release of antibiotics and bacteria into the environment, but it would be expensive; therefore, "risk-assessment" approaches tells us that we can save some money by setting loose regulations: whatever the (perceived) environmental and health consequences of this, would be cheaper that tightening the rules. Aside from the inability of risk-assessment approaches to actually assess the risks, this behavior is defining acceptable risks on terms of financial savings.

Antibiotics and antibiotic resistance in the environment mark one of the many convergences of public health and ecology; in the end, both deal with the wellbeing of living organisms. Free-market theologies have their focus and faith at precisely the other end of the scale. While it may be permissible for free-market to decide whether a brand of cell phones or cosmetics prevail or not, environmental and health regulations must be completely detached from it. This may sound unrealistic, but our very lives depend on understanding it, and acting accordingly.

REFERENCES

Allen, H. K., Looft, T., Bayles, D. O., Humphrey, S., Levine, U. Y., Alt, D. & Stanton, T. B. (2012) Antibiotics in feed induce prophages in swine fecal microbiomes. *mBio*, 2, 00260-11.

Aminov, R. I. (2009) The role of antibiotics and antibiotic resistance in nature. *Environ. Microbiol.*, 11, 2970–2988.

Aminov, R. I. & Mackie, R. I. (2007) Evolution and ecology of antibiotic resistance genes. *FEMS Microbiol. Lett.*, 271, 147–161.

Arumugam, M., Raes, J., Pelletier, E., Le Paslier, D., Yamada, T., Mende, D. R., Fernandes, G. R., Tap, J., Bruls, T., Batto, J. M., Bertalan, M., Borruel, N., Casellas, F., Fernandez, L., Gautier, L., Hansen, T., Hattori, M., Hayashi, K., Kleerebezem, M., Kurokawa, K., Leclerc, M., Levenez, F., Manichanh, C., Nielsen, H. B., Nielsen, T., Pons, N., Poulain, J., Qin, J., Sicheritz-Ponten, T., Tims, S., Torrents, D., Ugarte, E., Zoetendal, E. G., Wang, J., Guarner, F., Pedersen, O., De Vos, W. M., Brunak, S., Doré, J., Consortium, M., Weissenbach, J., Ehrlich, S. D. & Bork, P. (2011) Enterotypes of the human gut microbiome. *Nature*, 473, 174–180.

Baharoglu, Z. & Mazel, D. (2011) *Vibrio cholerae* triggers SOS and mutagenesis in response to a wide range of antibiotics: a route towards multiresistance. *Antimicrob. Agents Chemother.*, 55, 2438–2441.

Bengtsson-Palme, J. & Larsson, D. G. J. (2015) Antibiotic resistance genes in the environment: prioritizing risks. *Nat. Rev. Microbiol.*, 13, 396.

Bernier, S. P. & Surette, M. G. (2013) Concentration-dependent activity of antibiotics in natural environments. *Front. Microbiol.*, 4, 20.

Claesson, M. J., Jeffery, I. B., Conde, S., Power, S. E., O'Connor, E. M., Cusack, S., Harris, H. M. B., Coakley, M., Lakshminarayanan, B., O'Sullivan, O., Fitzgerald, G. F., Deane, J., O'Connor, M., Harnedy, N., O'Connor, K., O'Mahony, D., Van Sinderen, D., Wallace, M., Brennan, L., Stanton, C., Marchesi, J. R., Fitzgerald, A. P., Shanahan, F., Hill, C., Ross, R. P. & O'Toole, P. W. (2012) Gut microbiota composition correlates with diet and health in the elderly. *Nature*.

Claus, S. P., Ellero, S. L., Berger, B., Krause, L., Bruttin, A., Molina, J., Paris, A., Want, E. J., De Waziers, I., Cloarec, O., Richards, S. E., Wang, Y., Dumas, M. E., Ross, A., Rezzi, S., Kochhar, S., Van Bladeren, P., Lindon, J. C., Holmes, E. & Nicholson, J. K. (2011) Colonization-induced host-gut microbial metabolic interaction. *mBio*, 2, e00271-10.

Cohen, M. L. (2000) Changing patterns of infectious disease. *Nature*, 406, 762–767.

Diaz Heijtz, R., Wang, S., Anuar, F., Qian, Y., Björkholm, B., Samuelsson, A., Hibberd, M. L., Forssberg, H. & Pettersson, S. (2011) Normal gut microbiota modulates brain development and behavior. *Proc. Natl. Acad. Sci. USA.*

Finley, R. L., Collignon, P., Larsson, D. G. J., McEwen, S. A., Li, X. Z., Gaze, W. H., Reid-Smith, R., Timinouni, M., Graham, D. W. & Topp, E. (2013) The scourge of antibiotic resistance: the important role of the environment. *Clin. Infect. Dis.*, 57, 704–710.

Ford, T. (1994) Pollutant effects on the microbial ecosystem. *Environ. Health Perspect.*, 102 (suppl. 12), 45–48.

Forsberg, K. J., Patel, S., Gibson, M. K., Lauber, C. L., Knight, R., Fierer, N. & Dantas, G. (2014) Bacterial phylogeny structures soil resistomes across habitats. *Nature*, 509, 612–616.

Forsberg, K. J., Reyes, A., Wang, B., Selleck, E. M., Sommer, M. O. A. & Dantas, G. (2012) The shared antibiotic resistome of soil bacteria and human pathogens. *Science*, 337, 1107–1111.

Gareau, M. G., Sherman, P. M. & Walker, W. A. (2010) Probiotics and the gut microbiota in intestinal health and disease. *Nat. Rev. Gastroenterol. Hepatol.*, 7, 503–514.

Gibson, M. K., Forsberg, K. J. & Dantas, G. (2015) Improved annotation of antibiotic resistance determinants reveals microbial resistomes cluster by ecology. *ISME J.*, 9, 207–216.

Gilbert, P., McBain, A. & Lindsay, S. (2007) Biofilms, multi-resistance, and persistence. In Amábile-Cuevas, C. F. (Ed.) *Antimicrobial resistance in bacteria*. Wymondham, UK, Horizon Bioscience.

Gullberg, E., Albrecht, L. M., Karlsson, C., Sandegren, L. & Andersson, D. I. (2014) Selection of a multidrug resistance plasmid by sublethal levels of antibiotics and heavy metals. *mBio*, 5, e01918-14.

Heinemann, J. A. & Traavik, T. (2004) Problems in monitoring horizontal gene transfer in field trials of transgenic plants. *Nat. Biotechnol.*, 22, 1105–1109.

Heuer, H., Kopmann, C., Zimmerling, U., Krögerrecklenfort, E., Kleineidam, K., Schloter, M., Top, E. M. & Smalla, K. (2012) Effect of veterinary medicines introduced via manure into soil on the abundance and diversity of antibiotic resistance genes and their transferability. In Keen, P. L. & Montforts, M. H. M. M. (Eds.) *Antimicrobial resistance in the environment*. Hoboken, John Wiley & Sons.

Huang, Y. T. & Hsueh, P. R. (2010) Antimicrobial drug resistance in Asia. In Sosa, A. J., Byarugada, D. K., Amábile-Cuevas, C. F., Hsueh, P. R., Kariuki, S. & Okeke, I. N. (Eds.) *Antimicrobial resistance in developing countries*. New York, Springer.

Janzon, A., Kristiansson, E. & Larsson, D. G. J. (2012) Environmental microbial communities living under very high antibiotic selection pressure. In Keen, P. L. & Montforts, M. H. M. M. (Eds.) *Antimicrobial resistance in the environment*. Hoboken, John Wiley & Sons.

Kajander, E. O. & Çiftçioglu, N. (1998) Nanobacteria: an alternative mechanism for pathogenic intra- and extracellular calcification and stone formation. *Proc. Natl. Acad. Sci. USA*, 95, 8274–8279.

Keen, P. L. & Patrick, D. M. (2013) Tracking change: a look at the ecological footprint of antibiotics and antimicrobial resistance. *Antibiotics*, 2, 191–205.

Koike, S., Krapac, I. G., Oliver, H. D., Yannarell, A. C., Chee-Sanford, J. C., Aminov, R. I. & Mackie, R. I. (2007) Monitoring and source tracking of tetracycline resistance genes in lagoons and groundwater adjacent to swine production facilities over a 3-year period. *Appl. Environ. Microbiol.*, 73, 4813–4823.

Larsson, D. G. J. (2014) Antibiotics in the environment. *Upsala J. Med. Sci.*, 119, 108–112.

Li, Z., Sheridan, P. P. & Shields, M. S. (2015) The impact of enterohemorrhagic *Escherichia coli* (EHEC) on ciliate protozoan populations in municipal sewage. *Adv. Microbiol.*, 5, 668–676.

Liu, J., Prindle, A., Humphries, J., Gabalda-Sagarra, M., Asally, M., Lee, D. D., Ly, S., Garcia-Ojalvo, J. & Süel, G. M. (2015) Metabolic co-dependence gives rise to collective oscillations within biofilms. *Nature*, 523, 550–554.

Lupo, A., Coyne, S. & Berendonk, T. U. (2012) Origin and evolution of antibiotic resistance: the common mechanisms of emergence and spread in water bodies. *Front. Microbiol.*, 3, 18.

Martínez, J. L. (2012) Bottlenecks in the transferability of antibiotic resistance from natural ecosystems to human bacterial pathogens. *Front. Microbiol.*, 2, 265.

Martínez, J. L., Coque, T. M. & Baquero, F. (2015a) Prioritizing risks of antibiotic resistance genes in all metagenomes. *Nat. Rev. Microbiol.*, 13, 369.

Martínez, J. L., Coque, T. M. & Baquero, F. (2015b) What is a resistance gene? Ranking risk in resistomes. *Nat. Rev. Microbiol.*, 13, 116–123.

Martinez, J. L., Sánchez, M. B., Martínez-Solano, L., Hernandez, A., Garmendia, L., Fajardo, A. & Alvarez-Ortega, C. (2009) Functional role of bacterial multidrug efflux pumps in microbial natural ecosystems. *FEMS Microbiol. Rev.*, 33, 430–449.

Maslowski, K. M., Vieira, A. T., Ng, A., Kranich, J., Sierro, F., Yu, D., Schilter, H. C., Rolph, M. S., Mackay, F., Artis, D., Xavier, R. J., Teixeira, M. M. & Mackay, C. R. (2009) Regulation of inflammatory responses by gut microbiota and chemoattractant receptor GPR43. *Nature*, 461, 1282–1286.

McEwen, S. A. (2012) Human health importance of the use of antimicrobials in animals and its selection of antimicrobial resistance. In Keen, P. L. & Montforts, M. H. M. M. (Eds.) *Antimicrobial resistance in the environment*. Hoboken, John Wiley & Sons.

Mullany, P. (2014) Functional metagenomics for the investigation of antibiotic resistance. *Virulence*, 5, 443–447.

Oliveira, N. M., Martinez-Garcia, E., Xavier, J., Durham, W. M., Kolter, R., Kim, W. & Foster, K. R. (2015) Biofilm formation as a response to ecological competition. *PLoS Biol.*, 13, e1002191.

Ottman, N., Smidt, H., De Vos, W. M. & Belzer, C. (2012) The function of our microbiota: who is out there and what do they do? *Front. Microbiol.*, 2, 104.

Payne, A. N., Chassard, C., Zimmermann, M., Müller, P., Stinca, S. & Lacroix, C. (2011) The metabolic activity of gut microbiota in obese children is increased compared with normal-weight children and exhibits more exhaustive substrate utilization. *Nutr. Diabetes*, 1, e12.

Poole, K. (2012) Bacterial stress responses as determinants of antimicrobial resistance. *J. Antimicrob. Chemother.*, 67, 2069–2089.

Pruden, A., Larsson, D. G. J., Amézquita, A., Collignon, P., Brandt, K. K., Graham, D. W., Lazorchak, J. M., Suzuki, S., Silley, P., Snape, J. R., Topp, E., Zhang, T. & Zhu, Y. G. (2013) Management options for reducing the release of antibiotics and antibiotic resistance genes to the environment. *Environ. Health Perspect.*, 121, 878–885.

Qin, J., Li, Y., Cai, Z., Li, S., Zhu, J., Zhang, F., Liang, S., Zhang, W., Guan, Y., Shen, D., Peng, Y., Zhang, D., Jie, Z., Wu, W., Qin, Y., Xue, W., Li, J., Han, L., Lu, D., Wu, P., Dai, Y., Sun, X., Li, Z., Tang, A., Zhong, S., Li, X., Chen, W., Xu, R., Wang, M., Feng, Q., Gong, M., Yu, J., Zhang, Y., Zhang, M., Hansen, T., Sanchez, G., Raes, J., Falony, G., Okuda, S., Almeida, M., Lechatelier, E., Renault, P., Pons, N., Batto, J. M., Zhang, Z., Chen, H., Yang, R., Zheng, W., Li, S., Yang, H., Wang, J., Ehrlich, S. D., Nielsen, R., Pedersen, O., Kristiansen, K. & Wang, J. (2012) A metagenomic-wide association study of gut microbiota in type 2 diabetes. *Nature*, 490, 55–60.

Ruiz, G. M., Rawlings, T. K., Dobbs, F. C., Drake, L. A., Mullady, T., Huq, A. & Colwell, R. R. (2000) Global spread of microorganisms by ships. *Nature*, 408, 49–50.

Sardarabadi, H., Mashreghi, M., Jamialahmadi, K. & Dianat, T. (2014) Resistance of nanobacteria isolated from urinary and kidney stones to broad-spectrum antibiotics. *Iran. J. Microbiol.*, 6, 230–233.

Sørensen, S. J., Bailey, M., Hansen, L. H. & Wuertz, S. (2005) Studying plasmid horizontal transfer in situ: a critical review. *Nat. Rev. Microbiol.*, 3, 700–710.

Turnbaugh, P. J., Hamady, M., Yatsunenko, T., Cantarel, B. L., Duncan, A., Ley, R. E., Sogin, M. L., Jones, W. J., Roe, B. A., Affourtit, J. P., Egholm, M., Henrissat, B., Heath, A. C.,

Knight, R. & Gordon, J. I. (2009) A core gut microbiome in obese and lean twins. *Nature*, 457, 480–484.

Verraes, C., Van Boxstael, S., Van Meervenne, E., Van Coillie, E., Butaye, P., Catry, B., De Schaetzen, M. A., Van Huffel, X., Imberechts, H., Dierick, K., Daube, G., Saegerman, C., De Block, J., Dewulf, J. & Herman, L. (2013) Antimicrobial resistance in the food chain: a review. *Int. J. Environ. Res. Public Health*, 10, 2643–2669.

Xiong, W., Sun, Y., Ding, X., Wang, M. & Zeng, Z. (2015) Selective pressure of antibiotics on ARGs and bacterial communities in manure-polluted freshwater-sediment microcosms. *Front. Microbiol.*, 6, 194.

Xue, H., Cordero, O. X., Camas, F. M., Trimble, W., Meyer, F., Guglielmini, J., Rocha, E. P. C. & Polz, M. F. (2015) Eco-evolutionary dynamics of episomes among ecologically cohesive bacterial populations. *mBio*, 6, e00552-15.

Yim, G., Wang, H. H. & Davies, J. (2007) Antibiotics as signalling molecules. *Phil. Trans. R. Soc. B*, 362, 1195–1200.

Subject index

Milton Keynes UK
Ingram Content Group UK Ltd.
UKHW051855071024
449327UK00025B/1967